HORMONAL CONTROL OF TREE GROWTH

FORESTRY SCIENCES

Baas P, ed: New Perspectives in Wood Anatomy. 1982. ISBN 90-247-2526-7
Prins CFL, ed: Production, Marketing and Use of Finger-Jointed Sawnwood. 1982. ISBN 90-247-2569-0
Oldeman RAA, et al., eds: Tropical Hardwood Utilization: Practice and Prospects. 1982. ISBN 90-247-2581-X
Den Ouden P and Boom BK: Manual of Cultivated Conifers: Hardy in Cold and Warm-Temperate Zone. 1982. ISBN 90-247-2148-2
Bonga JM and Durzan DJ, eds: Tissue Culture in Forestry. 1982. ISBN 90-247-2660-3
Satoo T and Magwick HAI: Forest Biomass. 1982. ISBN 90-247-2710-3
Van Nao T, ed: Forest Fire Prevention and Control. 1982. ISBN 90-247-3050-3
Douglas J: A Re-appraisal of Forestry Development in Developing Countries. 1983. ISBN 90-247-2830-4
Gordon JC and Wheeler CT, eds: Biological Nitrogen Fixation in Forest Ecosystems: Foundations and Applications. 1983. ISBN 90-247-2849-5
Németh MV: The Virus-Mycoplasma and Rikettsia Disease of Fruit Trees. ISBN 90-247-2868-1
Duryea ML and Landis TD, eds: Forest Nursery Manual: Production of Bareroot Seedlings. 1984. ISBN 90-247-2913-0
Hummel FC, ed: Forest Policy: A Contribution to Resource Development. 1984. ISBN 90-247-2883-5
Manion PD, ed: Scleroderris Canker of Conifers. 1984. ISBN 90-247-2912-2
Duryea ML and Brown GN, eds: Seedling Physiology and Reforestation Success. 1984. ISBN 90-247-2949-1
Staaf KAG and Wiksten NA: Tree Harvesting Techniques. 1984. ISBN 90-247-2994-7
Boyd JD: Biophysical Control of Microfibril Orientation in Plant Cell Walls. 1985. ISBN 90-247-3101-1
Findlay WPK, ed: Preservation of Timber in the Tropics. 1985. ISBN 90-247-3112-7
Samset I: Winch and Cable Systems. 1985. ISBN 90-247-3205-0
Leary RA: Interaction Theory in Forest Ecology and Management. 1985. ISBN 90-247-3220-4
Gessel SP: Forest Site and Productivity. 1986. ISBN 90-247-3284-0
Hennessey TC, Dougherty PM, Kossuth SV and Johnson JD, eds: Stress Physiology and Forest Productivity. 1986. ISBN 90-247-3359-6
Shepherd KR: Plantation Silviculture. 1986. ISBN 90-247-3379-0
Sohlberg S and Sokolov VE, eds: Practical Application of Remote Sensing in Forestry. 1986. ISBN 90-247-3392-8
Bonga JM and Durzan DJ, eds: Cell and Tissue Culture in Forestry. Volume 1: General Principles and Biotechnology. 1987. ISBN 90-247-3430-4
Bonga JM and Durzan DJ, eds: Cell and Tissue Culture in Forestry. Volume 2: Specific Principles and Methods: Growth and Developments. 1987. ISBN 90-247-3431-2
Bonga JM and Durzan DJ, eds: Cell and Tissue Culture in Forestry. Volume 3: Case Histories: Gymnosperms, Angiosperms and Palms. 1987. ISBN 90-247-3432-0 (Set Vols. 1–3: ISBN 90-247-3433-9)
Richards EG, ed: Forestry and the Forest Industries: Past and Future. 1987. ISBN 90-247-3592-0
Kossuth SV and Ross SD, eds: Hormonal Control of Tree Growth. 1987. ISBN 90-247-3621-8

Hormonal Control of Tree Growth

Proceedings of the Physiology Working Group Technical Session, Society of American Foresters National Convention, Birmingham, Alabama, USA, October 6–9, 1986

edited by

SUSAN V. KOSSUTH
USDA Forest Service, University of Florida, Gainesville, Florida, USA

STEVE D. ROSS
Research Branch, British Columbia Ministry of Forests and Lands, Victoria, B.C., Canada

Reprinted from Plant Growth Regulation Volume 6, Nos 1–2, 1987
additionally containing an article by J.E. Webber and an index

1987 **MARTINUS NIJHOFF PUBLISHERS**
a member of the KLUWER ACADEMIC PUBLISHERS GROUP
DORDRECHT / BOSTON / LANCASTER

Distributors

for the United States and Canada: Kluwer Academic Publishers, P.O. Box 358, Accord Station, Hingham, MA 02018-0358, USA
for the UK and Ireland: Kluwer Academic Publishers, MTP Press Limited, Falcon House, Queen Square, Lancaster LA1 1RN, UK
for all other countries: Kluwer Academic Publishers Group, Distribution Center, P.O. Box 322, 3300 AH Dordrecht, The Netherlands

Library of Congress Cataloging in Publication Data

Hormonal control of tree growth.

(Forestry sciences)
Includes index.
1. Trees--Growth--Congresses. 2. Plant regulators--Congresses. 3. Plant hormones--Congresses. I. Kossuth, S. V. II. Ross, Steve D., 1941- . III. Society of American Foresters. Physiology Working Group. IV. Society of American Foresters. Convention (1986 : Birmingham, Ala.) V. Series.
SD396.H668 1987 634.9 87-24835

ISBN 90-247-3621-8

PRINTED IN THE NETHERLANDS

CONTENTS

HORMONAL CONTROL OF TREE GROWTH
PREFACE

This is the third annual compendium of a Technical Session of the Physiology Working Group of the Society of American Foresters held at the National Convention. Specialists in a dedicated area of tree physiology were invited to prepare chapter contributions synthesizing the status of knowledge in their area of expertise. Plant growth regulators (PGRs) was selected as the topic for in-depth examination at the 1986 Technical Session because a knowledge of how these "secondary messengers" regulate tree morphogenesis is vital to applications of biocontrol and biotechnology.

Plant growth regulators have been the subject of numerous reviews in recent years. However, few have dealt specifically with woody perennials, and they are generally confined to single processes and/or organs. This volume attempts to provide a more comprehensive treatise of PGRs as they influence various ontogenetic events in forest trees. Reproductive physiology, both sexual and asexual, is emphasized because of its relevance to current efforts directed at increasing efficiency in the breeding and production of genetically improved trees for reforestation. The chapters on vegetative growth will be of interest to silviculturists and urban foresters as they consider cultural treatments in the management of forests and individual trees for specific products and purposes. This book should serve as a valuable text and source of reference for students, researchers and other professionals interested in gaining a better understanding of PGRs.

The reader, however, who expects definitive answers to how PGRs function or can be used to control specific processes is likely to be disappointed. Lack of critical study, the complexity of PGR function in interaction with ontogeny and environment, and inadequacies of past bioassay and chromatographic techniques used to identify and quantify endogenous hormones are recurrent themes throughout this book. Consequently, our understanding of PGR function in forest trees is still quite rudimentary, though expanding rapidly as more laboratories are becoming equipped for more sophisticated analyses by means of immunoassays, mass spectrometry and nuclear magnetic resonance spectrometry. In preparing their chapters the authors were requested to evaluate the published literature to synthesize what is credibly known and to identify the directions that future research should take. This we believe they have done in a most readable and informative manner.

We wish to acknowledge the cooperation of the Society of American Foresters in creating for the Physiology Working Group an open forum at the convention for discussion of special topics bearing on land management practices. A special thanks is extended to Robert C. Biesterfeldt, Publications Management and Information Group Leader, Southeastern Forest Experiment Station, USDA Forest Service, Asheville, North Carolina, U.S.A. for editorial review in preparation of the manuscripts. We also wish to acknowledge the many scientists from whom the authors obtained peer review.

Susan V. Kossuth
Steve D. Ross

Plant Growth Regulation 6: 1–12

1. REJUVENATION OF FOREST TREES

M. S. GREENWOOD - Professor and Chair, Department of Forest Biology, College of Forest Resources, 122 Nutting Hall, University of Maine, Orono, ME 04469

ABSTRACT

Phase change, or maturation, in woody plants not only results in changes in growth behavior, but results in increased difficulty in vegetatively propagating select individuals. The process of maturation in woody plants is discussed, with emphasis on methods for reversal of the process, including use of plant growth regulators and tissue culture. Since maturation is manifested by changes in growth behavior, foliar morphology, phyllotaxy, reproductive competence, in addition to numerous other traits, care must be taken to evaluate the effectiveness of rejuvenation methods in terms of all of these traits. Several hypotheses on the mechanism for phase change are discussed.

Key Words: Maturation, phase change, woody plants, plant growth regulators, tissue culture.

1. INTRODUCTION

Maturation, or phase change in trees, is a developmental process which is at present poorly understood. It is exemplified by such characteristics as a reduced rate of growth, reduced efficiency in the rooting of cuttings, reduced ability to propagate mature plants in tissue culture, changes in foliar morphology, and the onset of flowering. Rejuvenation, or the reversal of maturation, is of interest for several reasons. First, the period of rapid growth associated with the juvenile phase is probably necessary to capture nonadditive (dominance) genetic variation when a superior individual is selected for clonal propagation. Second, the juvenile phase may be vital for the tree to compete with other vegetation soon after establishment (22). Third, in addition to difficulties in cloning mature individuals by tissue culture, recent observations that tissue culture plantlets derived from embryonic tissue of loblolly pine behave like mature trees (33) are cause for concern. Similar behavior has been reported for radiata pine (Smith, 1983, personal communication, Forestry Research Institute, Rotorua, New Zealand), and there are other reports of early flowering by tissue culture-derived plants (60). This "accelerated maturation" is especially paradoxical when compared with a number of other recent reports that tissue culture of mature tissue results in restoration of juvenile characteristics (14). This review examines the nature of the juvenile condition and critically evaluates some of the methods where rejuvenation is claimed. As a basis for this evaluation I will define the juvenile and mature states at least to a degree that one is not misled by incomplete or temporary restoration of juvenile characteristics or their premature loss. Maturation, as it is used in this review, means the ongoing process of phase change which results in relatively permanent developmental changes. The term is distinct from aging as used by Wareing (56) to describe changes due to increased size and complexity of the tree, which disappear when a scion is grafted onto young root stock. For example, orthotopic growth and some growth vigor is often restored after grafting a scion from a slow-growing lateral branch of a mature tree. The phenomenon of maturation, and more specifically, the role of plant growth regulators in the process have been recently and thoroughly reviewed (7, 24, 60). In this report, attention will be focused primarily on maturation in conifers due to their great economic importance and the resulting interest in cloning them *in vitro* (13).

Even though phase change in woody plants has been described frequently (8, 12, 49, 56), systematic attempts to carefully document the time course of maturation events are lacking. Zimmerman *et al*. (60) point out that reliance on the ability to flower as the major criterion for phase change may be too simplistic, since flowering can be temporarily induced on 2- to 3-month-old *Cupressus* seedlings by GA3, which otherwise appear totally juvenile, but may fail to occur on newly grafted mature scions because of root stock and environmental effects. Nonetheless, for the sake of this discussion, I would like to propose 4 developmental phases, each characterized by a unique set of morphogenetic competencies.

1.1. *Embryogenetic phase*. This phase includes development of a root-shoot axis from the zygote, resulting in a mature, dormant embryo within the seed. At present, tissue explants capable of somatic embryogenesis *in vitro* are limited mostly to embryonic tissue or to the female gametophyte. Nagmani and Bonga (37) have reported embryogenetically competent callus from the gametophytic tissue of *Larix decidua* seeds. Recently, Hakman *et al*. (25) reported somatic embryogenesis from isolated, immature embryos of developing Norway spruce (*Picea abies* Karst.) seeds; these results have been confirmed by Becwar *et al*. (3). More recently, embryogenic callus has also been obtained from mature zygotic embryos of Norway spruce (54). Gupta and Durzan (23) have described somatic embryogenesis from mature embryos of sugar pine (*Pinus lambertiana*, Dougl.). Merkle and Sommer (34) also reported isolation of embryogenetically competent callus from both immature and mature embryos of yellow poplar (*Liriodendron tulipifera* L.). Despite numerous attempts, somatic embryogenesis has proven more difficult from callus derived from parts of germinated seedlings or older trees, although Krogstrup (29) reports embryo-like structure derived from the cotyledons of 7-day-old Norway spruce seedlings.

1.2. *Seedling phase*. After germination of the seed, the seedlings of many conifers undergo a period of free growth (the stem units have not been pre-formed in a bud--see Sweet and Bollman [53]) and often produce juvenile-type foliage. In *Pinus*, *Larix* and certain other species, the first foliage produced by the epicotyl consists of acicular needles in the axils of which short shoots will eventually develop. Tissue explants from progressively older seedlings may show reduced ability to regenerate roots and shoots *in vitro* (5) but cuttings from seedlings in their first year generally root very easily (20 and Greenwood, unpublished data). Sexual reproductive structures are extremely rare during this phase, although isolated instances of strobili on pine seedlings less than a year old have been reported (51, 45, Ross and Pharis, 1987, this volume). In my opinion, this phase approximates an ideal juvenile stage. In *Pinus*, a convenient end point for this phase occurs when free growth ends and primordia for subsequent growth cycles or flushes are pre-formed in a bud.

1.3. Transition phase. Zimmerman (59) notes that young fruit trees may attain the competence to flower, but are unable to do so without promotive treatment. As development progresses, they flower increasingly well and eventually do so in their natural environment without any stimulatory treatment. Vegetative and growth characteristics also change gradually during this phase. For example, in loblolly pine, the capacity for height and diameter growth, as well as the number of branches produced per unit of stem length, decline over at least a 12-year period (22). In contrast, needle length and diameter as well as reproductive competence increased during this period. These results confirm those of an earlier study on radiata pine by Sweet (52) who noted declines for the same growth characteristics over a period of approximately 43 years. Many morphological traits change gradually with time, including branching habit, foliar morphology, pigmentation, and phyllotaxy, in addition to reproductive competence and growth rate (22, 44). Unfortunately, very few studies delineate the time course of these changes in woody species. Data from two such studies (on loblolly and radiata pines) are presented in Figures 1 and 2, which show a gradual increase in the ability of grafted scions from ortets of different ages to flower and a gradual decrease in the capacity for diameter growth as ortet age increases. Regenerative abilities by cuttings or tissue culture explants generally decline during this phase.

1.4. Mature phase. This phase is reached when reproductive competence is maximized, which occurs between age 15 and 29 years in radiata pine (see Fig. 1). Conversely, the capacity for height and diameter growth reach a minimum, and no further changes in foliar or branching characteristics are expected. Calli derived from mature tissue do not readily regenerate roots or shoots, and somatic embryogenesis has never been observed.

The success of methods which are supposed to return tissue explants from mature or transition phases to the seedling (juvenile) phase must be rigorously evaluated. This can be done by characterizing the 4 phases described above for each species in question, and comparing the developmental behavior of propagules which result from rejuvenation techniques with that of normal seedlings. For example, grafted scions from 1-year-old tissue culture plantlets which have been through a rejuvenation process should behave like scions from 1-year-old seedlings.

2. REJUVENATION METHODS

Numerous reports of production of propagules with seedling or juvenile phase characteristics from trees in the mature phase have been reported. The techniques employed include application of a cytokinin either during or immediately before explants are placed in culture, serial grafting, propagation of stump sprouts, or severe pruning. A comparison of some of these methods may provide insights into the maturation process, how it is controlled and whether or not it has been truly reversed.

Tissue Culture (Cytokinin Application): Rejuvenation of mature tissues *in vitro* usually involves application of a cytokinin (usually benzyla-

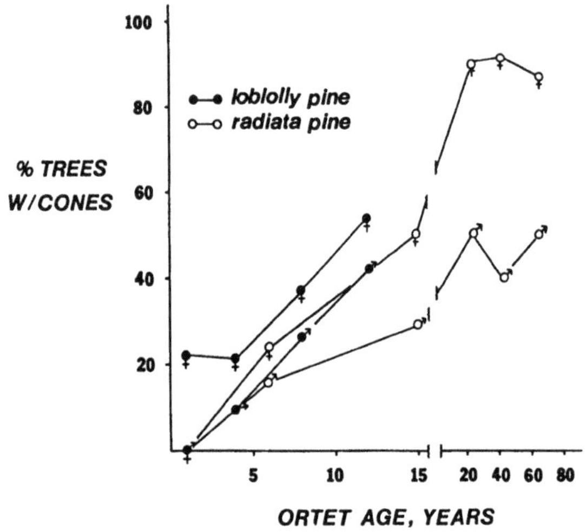

Figure 1. Female (♀), male (♂) strobilus production by grafted scions from different aged ortets of loblolly pine (♀,♂) and radiata pine (♀,♂). Strobilus counts made 3 years after grafting for loblolly pine, 4 years for radiata pine (22, 52). Scions from ortets at ages shown were grafted onto rootstock at the start of the experiment.

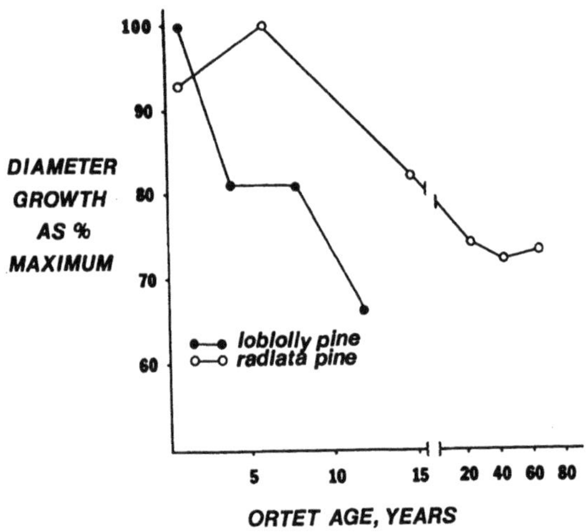

Figure 2. Diameter growth (as percent maximum) by grafted scions from different aged ortets from loblolly pine (●--●) and radiata pine (O--O). Measured 3 years after grafting for loblolly pine, 4 years for radiata pine (22, 52).

denine, BA) at some point to promote outgrowth of lateral buds or to promote the formation of adventitious shoots. One of the first examples of rejuvenation in vitro involved Cabernet Sauvignon, a grape cultivar used for the production of red wines for well over 1,000 years (36). Single node explants from newly emerged shoots produced shoots on medium without plant growth regulators, which were subsequently rooted. When these "initial plants" had produced 2-4 leaves, shoot tips with up to 2 unfolded leaves were subcultured, and shoots which arose from poorly-formed axillary buds on media with BA began to exhibit juvenile characteristics after 3-7 subcultures. Juvenile characteristics included lack of tendril production and a shift from distichous to spiral phyllotaxy. The rejuvenated state was maintained by serial culture for nearly 3 years even after production of more than 150 nodes, while rejuvenated (or seedling) plants planted into soil showed adult characteristics after production of 10-15 nodes. The authors conclude the distance between roots and shoots controls the expression of juvenile or mature characteristics.

Lyrene (32) reports rejuvenation of two varieties of blueberry also by serial subculture of shoots derived from mature plants. The mature shoots were started in liquid culture and later transferred to agar-containing medium, which increased the survival of mature explants. After subculturing shoot tips several times on solid media containing the cytokinin 2-ip (N^6 (Δ^2-isopentenyl) adenine), adventitious shoots with smaller leaves shorter internodes, thinner stems, which rooted and grew better than mature shoots were produced. Benzyladenine was ineffective in producing juvenile shoots (31). The author concludes that the juvenile shoots probably arose from single cells, which could be interpreted to mean that tissue culture conditions favored the proliferation of residual juvenile cells rather than actual rejuvenation. Alternatively, the process of adventitious shoot formation itself may be rejuvenative, since Bonga (7) concludes that all adventitious shoots are juvenile. Formation of axillary shoots in vitro may also be rejuvenative.

A similar approach was followed by Fouret et al. (14) to produce juvenile shoots from coastal redwood, starting with grafted scions from trees as old as 600 years. Shoot explants with 3 needles each were placed on solid medium with NAA (Napthaleneacetic acid) and BA, called a "milieu de multiplication" because additional shoots were produced. Every 3 weeks developing axillary shoots were transplanted to media without NAA or BA, allowed to elongate, and then rooted on a medium containing NAA. During transplantation to media, which encouraged outgrowth of roots and reactivation of the shoot, observations were made on the percentage of rooting, shoot elongation and phyllotaxy of the developing plantlets. With increasing exposure time to BA, shoots from mature trees became progressively (over a 17-week period) easier to root, exhibited more stem elongation, and produced longer leaves (arranged alternately on the stem), all considered to be juvenile characteristics. Explants from younger trees (1 and 50 years) resumed juvenile behavior more rapidly, and with fewer subcultures than those from the 600-year-old tree. Field trials of rejuvenated material with normal seedlings are not described.

There are numerous other examples of tissue culture propagation of explants from mature trees (4, 21, 41), with few details on how much rejuvenation has occurred, other than reports of juvenile-like foliage and increased ease of propagation.

In pines, application of cytokinin (usually BA) to mature material (with or without removal of the terminal shoot) forces the outgrowth of juvenile-like fascicular shoots (9, 28, 58). Plantlets can be produced from these fascicular shoots in vitro (17). When placed in culture, the BA-induced fascicular shoots from loblolly pine produce additional shoots with seedling-like foliage (1). With loblolly pine there was considerable tree-to-tree variation in the ability to form shoots in response to BA, but when present, they appeared progressively more juvenile with continued subculture. Plantlets rooted from these shoots were hardened off in the greenhouse and planted in the field. Although apparently normal in appearance, all (several thousand) the plantlets died during the following winter, while normal seedlings representing the same half sib families planted nearby survived (Carlson, W.C., 1985, personal communication, Weyerhaeuser Co., Hot Springs, AR, U.S.A. 71901). When presented with short days and cool temperatures, another group of plantlets of similar origin did set buds, which appeared small. During the induction of dormancy, no foliar reddening (anthocyanin production) was observed, which occurs in normal seedlings. Furthermore, in the course of laboratory freezing testing, the plantlets could only tolerate temperatures half as low as normal seedlings (Timmis, R. 1986, personal communication, Weyerhaeuser Co., Tacoma, WA, U.S.A.). For an unknown reason, these tissue culture-derived plants were not winter hardy which raises concern that they may have been altered by the tissue culture process.

Serial Grafting: Franclet (16) reports rejuvenation of Douglas-fir scions originating from a mature tree by repeatedly regrafting the scion onto juvenile rootstock. Gradual rejuvenation (based on foliar characteristics, growth rate, ease of grafting and ability of cuttings to root) is observed following each regraft. Scions from a 75-year-old Douglas fir grew almost as fast as the seedling control after 5 serial grafts. Serially grafted scions from a 95-year-old tree did exhibit more rapid growth but never reached the seedling control. Foliar characteristics of the mature scion also became more juvenile, but long-term observations on this material have not been reported.

Misson and Giot-Wirgot (35) report rejuvenation of western red cedar by micrografting onto rootstock grown in vitro (without any plant growth regulators). After grafting three times, scions from 10-year-old trees resumed production of juvenile-type foliage and elongated more rapidly. No further data are presented on persistence or completeness of this apparent rejuvenation.

Smeltzer (personal communication, International Paper Co., Bainbridge, GA, U.S.A., 1986) reports that serial regrafting caused rejuvenation in loblolly pine but was successful on only 2 of 10 clones after 3 serial grafts. Cuttings from the rejuvenated scions became much easier to root, and the resultant plants exhibited the juvenile characteristics described by Greenwood (22); i.e., smaller needles, yellower foliage, and decreased bud size. Field trials including normal seedlings and rooted cuttings from the rejuvenated scions have been recently established.

Hackett (24) points out that unless a control comparing mature scions grafted onto mature rootstock with similar scions on juvenile rootstock is established, it may be difficult to separate reversal of aging effects from actual rejuvenation. However, finding comparable juvenile and mature rootstock is difficult. Overall, long-term side-by-side comparisons of seedlings and rejuvenated plantlets of comparable size are needed to see if the time course of transition between phases 2-3 (described above) is the same for both.

Propagation Of Shoots Derived From Stump Sprouts Or Severe Pruning: Whether or not the juvenile appearance of stump sprouts or lateral shoots which emerge after pruning is the result of rejuvenation or outgrowth of vestigial pockets of juvenile tissue is at present unknown. The maintenance of juvenile characteristics in radiata pine and Norway spruce by hedging is well documented (6, 30, 46, 47). If new growth is removed (annually or biennially) from seedlings of these species, the fascicular or axillary shoots subsequently produced remain juvenile for many years, and provide a source for mass production of rooted cuttings which appear to behave like seedlings.

Some conifers respond to severe pruning or fire damage by producing juvenile-appearing shoots from branches or roots (15, 48), which are easier to root than mature shoots. Follow-up studies to demonstrate the completeness of the rejuvenation of these shoots are lacking. Franclet (17) reports that repeated pruning over a 7-year period of rooted cuttings taken from *Platanus acerifolia* trees planted in 1785 resulted in progressively more vigorous sprouts. Whether this observation represents actual rejuvenation as opposed to reversal of aging effects is difficult to determine from the data presented.

Hedera helix--A Special Case? Before discussing the validity and significance of the rejuvenation experiments just described, some discussion of phase change in English ivy (*Hedera helix* L.) is in order. The reason for discussing this work separately is that use of *Hedera* as a model system for other woody plants, especially trees, could be misleading. In *Hedera*, phase change only involves part of the plant so that distinctly separate juvenile and mature phases coexist on an individual. The mature growth usually appears to be associated with vines which have been able to grow vertically for some time (up a wall or tree trunk), and are infrequent on horizontally growing vines which may be the same age and size as those which have grown vertically. Bearing flowers well above the ground has obvious advantages with regard to avoidance of seed predation, facilitation of pollination and seed dispersal. In trees, although there may be some variation in the extent of phase change within a given tree (49), the periphery of the entire above-ground portions eventually become mature in varying degrees. Juvenile-like shoots only reappear in response to pruning or chemical treatment. In addition, the gibberellins (GAs) can induce rejuvenation in *Hedera* (42, 44), while inducing flowering but not necessarily maturation in conifers (60). Frydman and Wareing (18, 19) conclude that the juvenile form may have a somewhat higher concentration of GA than the mature form, and that this increase may be due to proximity to actively growing roots, which may produce GA.

Nonetheless, the morphological changes associated with phase change in *Hedera* are closely similar to those reported for trees (44). In addition to changes in growth rate, leaf morphology and pigmentation, phyllotaxy, rootability of cuttings and reproductive competence, grafted scions or rooted cuttings from the mature phase retain their mature characteristics. In addition, callus derived from juvenile tissue grows much faster than callus from mature tissue (38). However, the developmental competence of the mature callus contrasts sharply with that from trees, in that juvenile-appearing plantlets can be readily obtained via somatic embryogenesis (2) although their occurrence is erratic. Callus from mature trees, on the other hand, form shoots or roots reluctantly if at all, and somatic embryogenesis has never been reported.

While it is tempting to rely on *Hedera* as a model system for phase change, the possibility must be considered that maturation in trees, which appears to be harder to reverse, may have a different developmental basis. The possibility that maturation results from alteration of specific genes or changes in their activity must be considered. As yet, little is known about the role of genes in phase change, but some preliminary studies have been conducted on Hedera. Nuclear DNA content has been compared between juvenile and mature tissue with conflicting results. Schaffner and Nagl (50) report 71% more DNA in 2C nuclei in the mature phase, while Domoney and Timmis (11) report no difference. Polito and Chang (39) report no differences in nuclear morphology between the 2 phases, and Polito and Alliata (38) report no difference in DNA content per nucleus for juvenile and mature callus. Rogler and Dahmus (43) report only minor differences in RNA production by the two phases. They conclude that the rate of transcription of some RNA may differ between juvenile and mature tissue, but large changes in RNA transcribed from redundant DNA species were not found.

3. DISCUSSION

Numerous reports of woody plant rejuvenation have been discussed, and some, especially grape, blueberry and English ivy, appear at present to be somewhat convincing. While a zygote resulting from fusion of gametes that result from meiosis represents true rejuvenation, there is hope that some of the above examples approximate at least some of the process. However, the possibility must be considered that in some cases rejuvenation is mimicked due to genetic alteration and has not really occurred. Long-term comparison of the time course of maturation in normal seedlings and rejuvenated plantlets will at least partially settle this question. To completely answer it might require characterization of phase change at the molecular level to determine if indeed genes are altered during the process.

There is a good deal of recent evidence that somatic plant cells are capable of considerable and rapid genetic change, especially in the highly repeated parts of the genome (55). In flax, these changes can result from either hybridization, nutrient stress or tissue culture. Hughes (27) presents evidence that tobacco plants of tissue culture origin have developed a sterility barrier with wild type plants. While these observations raise concern about clonal fidelity in tissue culture, they may provide insights into the nature of the accelerated maturation mentioned in the introduction, phase change itself, and the role of cytokinin in these processes.

Root-produced cytokinins may also play a role in the apparent rejuvenation resulting from serial micrografting, severe pruning or outgrowth of stump sprouts. Root-produced gibberellins might also be expected to affect rejuvenation, since Zimmerman et al. (60) claim that GA's may be the most important hormone involved in maturation. But, with some exceptions, their application does not result in rejuvenation of woody plants. Although most of the important plant growth regulators have been reported in root extracts, xylem exudates and phloem sap, it is technically very difficult to show that they are synthesized in the roots and transported to the shoot (45). Roots have, nonetheless, been implicated as a possible source of cytokinins which are exported to the shoot, where they affect development (10, 57).

If indeed roots export cytokinins, the proximity of a scion during serial micrografting, or a stump sprout to the root system may present an opportunity for exposure to relatively high concentrations of root-produced cytokinin.

A major question remains concerning the developmental level at which rejuvenation occurs. Hackett (24) discusses whether the stability of the mature phase is determined at the level of the individual cell, the entire apical meristem, or is due to correlative effects involving the whole plant. At present there is equivocal evidence supporting all three possibilities, and it is not yet possible to pinpoint where phase change is determined. How does one account for the stable behavior of even extremely tiny grafted mature scions, which will, nonetheless, exhibit juvenile characteristics after serial grafting or tissue culture? In Hedera, mature callus cultures retain slower growth rates, but can give rise to apparently juvenile plants via somatic embryogenesis. Thus, mature behavior is not completely stable, and under the right conditions will revert to a more juvenile-like condition. This developmental plasticity can be accounted for if mature cells themselves are reprogrammed to somehow become juvenile, or if both juvenile and mature cell populations coexist within the apex. Maturation could then result from an increase in the proportion of mature cells in the apical meristem. Conversely, rejuvenation by tissue culture or serial micrografting could promote relatively more rapid division of vestigial juvenile cells, which would gradually increase in number relative to the mature population. Thus, phase change could simultaneously have a cellular basis (where some cells are irreversibly determined as mature), but explants could respond to changed external stimuli (like BA), so that remaining undetermined juvenile cells could be stimulated to divide more actively. A gradual and continual reduction in the proportion of juvenile cells with age might explain why explants from progressively older woody plants respond less and less well to rejuvenation treatments (14, 17).

Loss of developmental totipotency is an indication of cellular determination in both plants and animals. In plants, this process may be less irreversible, but in both cases, the nature of this phenomenon is not understood. If phase change has a genetic basis, then probing the genomes of both mature and juvenile cells using recombinant DNA techniques may detect some significant differences. Even if no unique polyA RNA's are found associated with the juvenile and mature states, cDNA probes could be developed to look for genetic polymorphisms or other genetic alterations, like transposable genetic elements, that may be associated with phase change. Methylation of particular DNA sequences may also be involved in the control of development (40), and Hood and Libby (26) have proposed that this process may be involved in phase change. Investigation of phase change at the molecular level is needed to assess these and other hypotheses.

In Hedera and most woody plants, the transition from the juvenile to mature phase occurs gradually, but the numerous traits associated with the mature phenotype do not all appear at the same time (8, 44). Thus, care must be exercised in viewing maturation as a single process; instead, it may be a composite of several traits changing independently.

Acknowledgments: I would like to thank Henry Amerson, Wesley Hackett, Keith Hutchison and John Smagula for very helpful reviews of this manuscript.

LITERATURE CITED

1. ABO EL-NIL MM 1982 Method for asexual reproduction of coniferous trees. Patent No. 4,353,184. October 12. U.S. Patent Office.
2. BANKS MS 1979 Plant regeneration from callus from two growth phases of English ivy, *Hedera helix* L. Z. Pflanzenphysiol. 92: 349-353.

3. BECWAR MR, TL NOLAND, SR WANN 1986 Somatic embryo development and plant regeneration from embryogenic Norway spruce callus. In: Recent developments in forest biotechnology. Proc. TAPPI Conf., in press.
4. BEKKAOUI F, A FRANCLET, N WALKER 1985 Culture in vitro de meristemes de Douglas age et juvenile. Annales Afocel 1984, pp. 45-73.
5. BERLYN GP, RE BECK 1980 Tissue culture as a technique for studying meristematic activity. In: Control of shoot growth in forest trees, Proc. IUFRO Symposium, June 20-24, Fredericton, NB, Can., CHA Little ed. pp. 305-324.
6. BOLSTAD PV, WJ Libby 1982 Comparison of radiata pine cuttings of hedge and tree-farm origin after growing seasons. Silvae Genetica 31:9-13.
7. BONGA JM 1982 Vegetative propagation in relation to juvenility, maturity, and rejuvenation. In: Tissue culture in forestry. JM Bonga, DJ Durzan eds., Martinus Nijhoff/Dr W. Junk, Publishers, pp. 387-412.
8. BORCHERT R 1976 Differences in shoot growth patterns between juvenile and adult trees and their interpretation based on systems analysis of trees. Acta Hort. 56:123-130.
9. BOURIQUET R, M TSOGAS, A BLASELLE 1985 Essais de rejeunissement de l'épicéa par les cytokinins. Annales Afocel 1984, pp. 173-185.
10. CARMI A, J VAN STADEN 1983 Role of roots in regulating the growth rate and cytokinin content in leaves. Plant Physiol. 73:76-78.
11. DOMONEY C, JN TIMMIS 1980 Ribosomal RNA gene redundancy in juvenile and mature ivy (Hedera helix). J. Expt. Bot. 31:1093-1110.
12. DOORENBOS J 1965 Juvenile and adult phases in woody plants. In: W Ruhlond ed., Encyl. Plant Physiol. Vol. XV/1, pp. 1222-1235. Springer, Berlin.
13. FARNUM P, R TIMMIS, JL KULP 1983 Biotechnology of forest yield. Science 219:694-702
14. FOURET Y, Y ARNAUD, C LARRIEU 1984 Rajeunissement in vitro sequoia sempervirens-effect du nombre et de la fréquence des repiquages-recherche de criteres proceses de juvénilité. Annales Afocel 1984, pp. 111-137.
15. FRANCLET A 1979 Rajeunissement des arbres adultes in vue de leur propagation vegetative. Afocel No. 12, Micropropagation D'Arbres Forestiers, pp. 3-18.
16. FRANCLET A 1981 Rajeunissement et propagation végtétative de lignieux. Annales Afocel 1980, pp. 12-39.
17. FRANCLET A, A DAVID, H DAVID, M BOULAY 1980 Premièr mis en évidence morphologique d'un rejeunissemet de méristèmes primares caulinaires de Pin maritime âgé (Pinus pinaster Sol.). C. R. Acad. Sc. Paris, 290, Serie D:927-930.
18. FRYDMAN VM, PF WAREING 1973 Phase change in Hedera helix L. I. Gibberellin-like substances in the two growth phases. J. Exp. Bot. 24:1131-1138.
19. FRYDMAN VM, PF WAREING 1973 Phase changes in Hedera helix L. II. The possible role of roots as a source of shoot gibberellin-like substances. J. Expt. Bot. 24:1139-1148.
20. GARDNER FE 1929 The relationship between tree age and the rooting of cuttings. Proc. Am. Soc. Hort. Sci. 26:101-104.

21. GOUBLAY DE NANTOIS T DE LA 1980 Rajeunissement chez le Douglas (*Pseudotsuga menziesii*) en vue de la propagation vegetative. In: Memoire l'Universite Pierre et Marie Curie, Paris VI, p. 44.
22. GREENWOOD MS 1984 Phase change in loblolly pine: shoot development as a function of age. Phys. Plant. 61:518-522.
23. GUPTA PK, DJ DURZAN 1986 Somatic polyembryogenesis from callus of mature sugar pine embryos. Bio. Tech. 4:643-645.
24. HACKETT WP 1985 Juvenility, maturation, and rejuvenation in woody plants. Hort. Reviews 7:109-155.
25. HAKMAN I, LC FOWKE, S vonARNOLD, T ERIKSSON 1985 The devevelopment of somatic embryos in tissue cultures initiated from immature embryos of *Picea abies* (Norway spruce). Plant Sci. 38:53-59.
26. HOOD JV, WJ LIBBY 1978 Continuing effects of maturation state in radiata pine and a general maturation model. *In*: Propagation of higher plants through tissue culture. KW Hughes, R Henke, M Constantin, eds. Proc. of Symposium at Univ. of Tennessee, Knoxville, TN, April 16-18.
27. HUGHES KW 1986 Tissue culture derived crossing barriers. Amer. J. Bot. 73:323-329.
28. KOSSUTH SV 1978 Induction of fascicular bud development in *Pinus sylvestris* L. HortSci. 13:174-176.
29. KROGSTRUP P 1986 Embryolike structures from cotyledons of Norway spruce (*Picea abies*). Can. J. For. Res. 16:664-668.
30. LIBBY WJ 1977 Rooted cuttings in production forests. *In*: Proc. 14th S. For. Tree Imp. Conf., Publication No. 36. Eastern Tree Seed Lab, P.O. Box 819, Macon, GA 31202, U.S.A. pp. 13-19.
31. LYRENE PM 1980 Micropropagation of rabbit eye blueberries. HortScience 15:80-81.
32. LYRENE PM 1981 Juvenility and production of fast-rooting cuttings from blueberry shoot cultures. J. Amer. Hort. Soc. 106:396-398.
33. McKEAND SE 1986 Expression of mature characteristics by tissue culture plantlets derived from embryos of loblolly pine. J. Amer. Soc. Hort. Sci. 111(4):619-623.
34. MERKLE SA, ME SOMMER 1986 Somatic embryogenesis in tissue cultures of *Liriodendron tulipifera* L. Can. J. For. Res. In press.
35. MISSON JP, P GIOT-WIRGOT 1985 Rajeunissement d'un clone de thuja en vue de sa multiplication *in vitro*. Annales Afocel 1984. pp. 187-197.
36. MULLINS MG, Y NAIR, P SIMPET 1979 Rejuvenation *in vitro*: induction of juvenile characters in an adult clone of *Vitis vinifera* L. Ann. Bot. 44:623-627.
37. NAGAMANI R, JM BONGA 1985 Embryogenesis in subcultured callus of *Larix decidua*. Can. J. For. Res. 15:1088-1091.
38. POLITO VS, V ALLIATA 1981 Growth of calluses derived from shoot apical meristems of adult and juvenile English ivy (*Hedera helix* L.). Plant Sci. Let. 22:387-393.
39. POLITO VS, YC CHANG 1984 Quantitative nuclear cytology of English ivy (*Hedera helix* L.). Plant Sci. Let. 34:369-377.
40. RAZIN A, M SZYF 1984 DNA methylation patterns: formation and function. Biochem. and Biophys. Acta 782:331-342.
41. RIFFAUD JL, D CORNU 1981 Utilisation de la culture in vitro pour la multiplication de merisiers adultes (*Prunus avium* L.) sélectionnés en forêt. Agronomie 8:633-640.
42. ROBBINS WJ 1960 Further observations on juvenile and adult *Hedera helix*. Amer. J. Bot. 47:485-491.

43. ROGLER CE, ME DAHMUS 1974 Gibberellic acid-induced phase change in Hedera helix as studied by deoxyribonucleic acid-ribonucleic acid hybridization. Plant Physiol. 54:88-94.
44. ROGLER CE, WP HACKETT 1975 Phase change in Hedera helix: Induction of the mature to juvenile phase change by gibberellin A3. Physiol. Plant. 34:141-147.
45. ROSS SD, RP PHARIS, W BINDER 1983 Growth regulators and conifers: their physiology and potential uses in forestry. In: LG Nickell ed., Plant Growth Regulating Chemicals, Vol. II, CRC, Boca Raton, FL, pp. 35-78.
46. ROULUND M 1974 Comparative study of characteristics of seedlings and clonal cuttings. N.Z. J. For. Sci. 4: 78-86.
47. ST CLAIR, JB, J KLEINSCHMIT, J SVOLBA 1985 Juvenility and serial vegetative propagation of Norway spruce clones (Picea abies Karst.). Silvae Gen. 42-48.
48. SANTAMOUR FS 1965 Rooting of pitch pine stump sprouts. Tree Planters' Notes 70:7-8.
49. SHAFFALITZKY DE MUCKADELL M 1959 Investigations of aging of apical meristems in woody plants and its importance in silviculture. Forstl. Forsogsv. Danm. 25, 310-455.
50. SCHAFFNER KJ, W NAGL 1979 Differential DNA replication involved in transition from juvenile to adult phase in Hedera helix Araliaceae). Pl. Syst. Evol. Suppl. 2:105-109.
51. SMITH WH, RN KONAR 1969 Initiation of ovulate strobili in cotyledon-stage seedlings of Pinus elliotii. Can. J. Bot. 47:624-626.
52. SWEET GB 1973 The effect of maturation on the growth and form of vegetative propagules of Radinta pine. N.Z. Jour. For. Sci. 3:191-210.
53. SWEET GB, MP BOLLMANN 1976 The terminology of pine shoot growth. N.Z. Jour. For. Sci. 6:393-396.
54. VON ARNOLD S, I HAKMAN 1986 Effect of sucrose on initiation of embryogenic callus cultures from mature zygotic embryos of Picea abies (L.) Karst. (Norway spruce). J. Plant Physiol. 122:261-265.
55. WALBOT V, CA CULLIS 1985 Rapid genomic change in higher plants. Ann. Rev. Plant Physiol. 36:367-396.
56. WAREING PF 1959 Problems of juvenility and flowering in trees. J. Linn. Soc. Lond. Bot. 56:282-289.
57. WAREING PF 1980 Root hormones and shoot growth. In: Control of shoot growth in trees. CHA Little ed., Proc. IUFRO Workshop, July 20-24, Fredericton, N.B., pp. 237-256.
58. WHITEHILL SJ, WW SCHWABE 1975 Vegetative propagation of Pinus sylvestris. Physiol. Plant. 35:66-71.
59. ZIMMERMAN RH 1973 Juvenility and flowering of fruit trees. Acta Hort. 34:139-142.
60. ZIMMERMAN RH, WP HACKETT, RP PHARIS 1985 Hormonal aspects of phase change and precocious flowering. In: Encyclopedia of Plant Physiology, RP Pharis, DM Reid eds., Springer-Verlag, Heidelberg, vol. 11(N.S.), pp. 79-115.

Plant Growth Regulation 6: 13–35

2. THE ROLE OF PLANT GROWTH REGULATORS IN PROMOTION OF FLOWERING

M. BONNET-MASIMBERT AND J. B. ZAERR

I.N.R.A. Station d'Amelioration des Arbres Forestiers, Centre de Recherches d'Orleans, Ardon, 45160 Olivet, France and Professor, Dept. of Forest Science, Oregon State University, Corvallis, Oregon 97331

ABSTRACT. Research on flowering in forest trees has pursued two main objectives: 1) developing methods for stimulating flowering of selected genotypes to accelerate progeny testing, and 2) hastening the production of genetically-improved seeds in seed orchards. Cultural techniques have been used to some extent in the past for these purposes, but the use of plant-growth regulators (PGRs), particularly the gibberellins (GAs), has been increasingly successful. GAs have been shown to induce flowering in most conifers, but GA application on an operational basis has been worked out for only a few species. GAs have been applied effectively by topical treatment, by stem or branch injections, and by foliar sprays. The timing and concentration of applied chemicals are important and vary from species to species. Another PGR, naphthalene-acetic acid, is synergistic with GAs in some species. Adjunct cultural treatments are often used in combination with treatment with GAs. The success of treatment with GAs is determined to some extent upon the stage of development of the treated plants. The endogenous PGRs regulating flowering are not completely known but it appears that the GAs, the cytokinins, and abscisic acid may all be changed by treatments which induce flowering. Improved techniques becoming available for measuring PGRs are beginning to stimulate research on the endogenous PGRs. The PGRs will continue to play a major role in stimulating flowering in forest trees.

1. INTRODUCTION

The ability of trees to flower is not a selection criterion used by tree breeders who are mainly interested in increasing wood production and improving wood quality. But flowering is necessary for breeding and there are many forest geneticists and physiologists investigating the physiology of flowering in forest trees. To date this research effort has been directed almost exclusively toward conifers.

Flowering in forest trees has generally been of little concern to foresters. On the other hand, fruit trees have a long history of clonal selection for increasing fruit production and improving quality. Breeders of forest trees are mainly interested in increasing wood production and improving wood quality rather than flowering. It is clear, however, that flowering is necessary for conventional breeding. Furthermore, instead of the few clonal varieties used for fruit trees, a large number of clones are necessary for forest seed orchards in order to maintain sufficient variability in the seeds to be produced.

Forest geneticists and physiologists investigating the flowering in trees face two kinds of problems: juvenility and irregularity of flowering. Juvenility is that period during the early life of a tree when flowering normally cannot be induced under otherwise favorable conditions, the so-called "ripeness to flower" period. But even when the trees are supposedly mature, they may not flower every year because of the absence of the "opportunity to flower" [44]. Lack of opportunity may be due to unfavorable climatic or other environmental factors, including competition between trees.

The objectives of flowering research, therefore, generally follow one of two lines: 1) how to stimulate the flowering of selected genotypes as early as possible in order to accelerate testing their progeny, and 2) how to hasten and regulate the production of genetically improved seeds in seed orchards. The main difference between these two objectives might be the degree of sophistication acceptable for the treatment, considering the number of trees to be treated, and the long-term effect of those treatments on the survival of treated trees. However, recent developments with potted seed orchards as opposed to conventional soil-based orchards may reduce these differences [92].

According to Ross and Pharis [95], basic treatments that can be used to promote flowering in forest trees are: water stress, root pruning, girdling, application of high temperature, nitrogen fertilizers, or appropriate growth regulators. In the present paper we emphasize how plant growth regulators (PGRs) can be used, alone or in combination with other kinds of treatment, to stimulate flowering. We will also touch upon some possible roles of the several PGRs.

2. STIMULATION OF FLOWERING THROUGH THE USE OF PLANT GROWTH REGULATORS

2.1. Background: the role of gibberellins (GAs)

Gibberellic acid (GA_3) was first demonstrated to induce flowering in members of the Cupressaceae and Taxodiaceae over 20 years ago [34, 68]. But it was just 10 years ago that significant results were obtained with species in the Pinaceae family, which are important for forestry in the temperate zone. It was demonstrated that GAs, especially the relatively less polar ones, were highly effective in stimulating flowering of lodgepole pine (*Pinus contorta* Dougl. ex Lowd.) [75] and Douglas-fir (*Pseudotsuga menziesii* (Mirb.) Franco) [64]. These results stimulated flowering research which, up to that time, had been limited to traditional cultural practices such as the application of water stress, root pruning, nitrogen fertilizer, and increase of temperature. But these cultural treatments produced only sporadic results at best [32, 82].

Detailed lists of species on which GAs have been shown to induce flowering are given in Pharis and Kuo [67] and Longman *et al.* [45] for species in the Cupressaceae and Taxodiaceae families and in Pharis and Ross [70, 95] for species of the Pinaceae. These reviews list almost 100 research papers dealing with the genera *Chamaecyparis*, *Cupressus*, *Juniperus*, *Libocedrus*, *Thuja*, *Thujopsis*, *Tetraclinis*, *Cryptomeria*, *Cunninghamia*, *Glyptostrobus*, *Metasequoia*, *Sequoia*, *Sequoiadendron*, *Taxodium*, *Larix*, *Picea*, *Pinus*, *Pseudotsuga*, and *Tsuga*.

Two-year-old potted seedlings of western hemlock (*Tsuga heterophylla* (Raf.) Sarg.), whose juvenile phase may persist for 20 to 25 years, were induced to flower by foliar application of a $GA_{4/7}$ mixture and water stress [9]. Spectacular results such as these and the fact that a flowering response has been shown in more than 40 species means that the effect of GAs on flowering is a general phenomenon, even if all of the physiological explanations remain unclear.

Also, it is important to note that seed cones that have been induced by GAs generally follow a normal development and set sound seeds. Sato [99] reported that seeds from GA-induced cones of *Cryptomeria*, *Chamaecyparis*, and *Metasequoia* weighed less than control seeds but had equal germination rates. Also, Owens and Molder [60] obtained 97% germination of seeds in a detailed study on the development of GA_3-induced cones. Puritch *et al.* [83] observed a significant reduction of the number of filled seeds per cone after GA treatment in some Douglas-fir seed orchards, but this reduction was more than offset by the increase in percentage of trees bearing cones and in the number of cones per tree [see also 96].

Thus, it appears that treatment of young material with GAs accelerates the normal course of development with respect to flowering without altering the juvenile phase status; i.e., GAs stimulate flower production during the juvenile phase [41]. Since it is often necessary to repeat GA treatment in order to continue flowering in subsequent years, GA treatment results in a temporary switch towards sexual development followed by normal growth development.

2.2. Methods and concentrations used for treatment with GAs

The primary consideration in applying PGRs to induce flowering is causing the desired amount of PGR to be taken up by the plant at a timing which will allow some of the potentially vegetative buds to be converted to sexual buds.

2.2.1. Topical application. Instead of treating the entire tree or branch, GAs may be applied to the expanding bud or shoot, close to the region where meristem differentiation occurs. This method is convenient, especially for preliminary studies, because different treatments (composition and concentrations) can be applied to different branches of one tree. Several techniques for topical application have been used:

1) Concentrated ethanol/aqueous solutions of GAs may be impregnated directly into the shoot in doses varying from 100 to 400ug per branch, in 10 to 20 ul droplets (e.g., on lodgepole pine [79], loblolly pine (*Pinus taeda* L.) [93], Douglas-fir [83], and Norway spruce (*Picea abies* (L.) Karst.) [21]).

2) Injection in ethanol/water solution underneath flaps of bark at the base of the shoot (e.g., on Sitka spruce (*Picea sitchensis* (Bong.) Carr.) [103]).

3) Application of a paste mixture with GAs and lanolin or sodium carboxy methyl cellulose into small incisions at the base of the

branch (e.g., on *Pinus thunbergii* Franco and *P. densiflora* Sieb. & Zucc., 20 mg GAs per branch [33]).

Although the topical method is useful for research purposes, it is not suitable for large-scale operational uses.

2.2.2. Stem injections. GAs can be injected directly into branches or stems to induce flowering. In contrast to sprays, this method allows a more accurate determination of the amount of GAs that penetrates the stem or branch, and is a useful technique for research purposes [22, 107].

An interesting approach used by Pharis and Ross for Douglas-fir [69], is the "hanging bottle" system. A hypodermic needle or tube is tightly fitted into a hole drilled in the xylem and fed with an aqueous solution of GAs from an hanging bottle. The solution is slowly and regularly drawn into the transpiration stream during the time of differentiation of meristems, an event which is not easy to determine precisely and which may vary for different parts of the tree. If the hole becomes occluded, which commonly occurs in about one week in Douglas-fir, a new hole can be drilled perpendicular to the first one, and above it. Moreover, it is possible to check the absorbed volume regularly. Similar to this method is the "cut-twig" or "cut-branch" system used by McMullan [55] on Douglas-fir. A lateral twig or branch is cut, the bark is peeled back from the cut end and a tube connected to a syringe or a bottle is tightly attached (see also [28] on loblolly pine). For research purposes and in intensively managed breeding orchards these methods are very useful, but technical problems are associated with their operational applications.

The most commonly used concentrations of GAs vary from 50 to 100 mg/l in 0.5 to 5% ethanol. Sometimes, at least on Douglas-fir, the 1-year- and 2-year-old needles are damaged by this treatment, but the expanding shoots are rarely damaged. The phytotoxicity may be associated with higher concentrations (5%) of ethanol rather than $GA_{4/7}$ *per se* [87]. This technique also gives good results on western hemlock [9, 97]; and larches (*Larix* sp.) [4], but it is unsatisfactory for pines or some spruces because resin blocks the tubing system. For these species GA solutions in alcohol are added directly to a hole drilled in the stem, e.g., 10 ul containing 0.5mg $GA_{4/7}$ for 2-year-old cuttings of lodgepole pine [43] and 1 to 2.5ml containing 100 to 250mg $GA_{4/7}$ for 14-year-old grafts of Sitka spruce [78]. This treatment also gives good results on Norway spruce [7] and Douglas-fir (Bower and Ross, personal communication, 1986). Philipson [78] considers this technique very effective for producing large quantities of pollen and seed cones on large field-grown trees. Positive results have also been obtained on Cupressaceae species [42].

2.2.3. Foliar sprays. Foliar spraying is the simplest method of applying GAs, but under field conditions rain and wind may reduce its efficacy considerably. In addition, because the cuticle layer on needles and shoots slows the absorption of aqueous solutions, much of the applied chemical may be wasted. Sprays are usually applied until the solution runs off the foliage (the "drip point"). Surfactants, which have a strong effect on the flowering response, are always necessary. GA_3 (100 to 500mg/l) applied with 0.05% "Tween 20" gave good results on *Cupressus*, *Cryptomeria*, and *Chamaecyparis* species [3, 34, 57, 58, 59, 73]. On

Pinaceae species, the first significant results were obtained when the cationic surfactant "Aromox-C12" (Armak Industrial Chemical, Chicago, IL.) (0.05% to 0.1% active ingredient) was used with $GA_{4/7}$ at concentrations ranging from 200 to 600 mg/l [88]. Positive results have subsequently been obtained by this method on Norway spruce [48], Scots pine (*Pinus sylvestris* L.) [14], lodgepole pine [111], western hemlock [9, 97], larch species [4], jack pine (*Pinus banksiana* Lamb.) [11], white spruce (*Picea glauca* (Moench) Voss) [12], and Douglas-fir [6, 88]. Repeated treatments (every week or every other week for at least 1 month) are usually necessary. Ethanolic sprays (50% ethanol) have also been successful for GA treatments in Norway spruce and Scots pine [50].

Ultra-low volume (ULV) spray equipment offers new opportunities for spray applications. In a recent test on Douglas-fir, Bower and Ross [8] found that the ULV sprays gave better flowering at 1/8 the $GA_{4/7}$ dosage applied by conventional mist sprayers. Addition of a spray oil ("Sunspray 6E", 2% v/v) improved the results. Optimum concentrations have yet to be determined for western hemlock, but due to the constraints of no wind and no rain at the time of application, the authors suggest that ULV sprays are best suited for indoor potted orchards [92].

2.2.4. GA concentration and the flowering response. The optimal concentration of GA is difficult to ascertain because of the large number of parameters to be measured, the wide range of concentrations that have been used, the occasional addition of other PGRs, and the paucity of papers reporting appropriate results [21, 28, 64, 72]. In addition, the variation between clones and even between ramets within a clone are considerable, making evaluation of dose/response relationships difficult. Statistically significant differences between treatments are difficult to obtain even though differences between means are large. Nevertheless, dose-response differences have been reported for western hemlock [97]. Greenwood [28] reports a dose-response curve for the effect of $GA_{4/7}$ on female flowering in loblolly pine which appears to be logarithmic and Ross [89] found a similar relationship in Douglas-fir. Although Luukkanen [49] found GA_4 to be slightly superior to GA_7 for Scots pine, the roles of GA_4 and GA_7 are not known; they could be equally effective or they could be considerably different in their effectiveness.

For physiological experiments, a more uniform experimental material would be desirable. Cuttings or plants reproduced in vitro culture would be more appropriate material [43] than the grafts that have been used in most studies.

2.3. The possible use of other growth regulators

GAs are the only PGRs commonly used for induction of precocious flowering and enhancement of flowering in conifers, but other PGRs have been shown to be useful, and some have shown a synergistic response with GAs [70, 95]. When the synthetic auxin napthalene-acetic acid (NAA) is applied alone, it is generally inefficient in promoting flowering, although it has been reported to have an effect on *Larix leptolepis* Gordon [29]. When added to GAs, NAA usually enhances the flowering response. Very clear positive results have been obtained with Douglas-fir [71, 72, 86], Sitka spruce [103], lodgepole pine [111] and white spruce [74]. For these species NAA is often included in the routine treatments with GAs, usually at one-tenth the concentration of the GAs. However, no response to NAA was found in western hemlock [97], jack pine [10], or larch sp. [4]. NAA reduced the number of male flowers induced by GAs in the hybrid *Cupressocyparis laylandii* [42].

The growth retardant cycocel CCC (2000 mg/l) was applied to Norway spruce by Bleymüller [2] and Chalupka [13] 2 weeks or more after treatment with GA_3. Flowering was enhanced and the effect seemed to depend upon the period between the two treatments.

Many of the adjunct treatments commonly applied, i.e., water stress, bark girdling that causes tissue wounding, high or low temperature, and flooding are able to produce "stress ethylene", a mechanism that involves 1-amino-cyclopropyl-1-carboxylic acid (ACC) synthase [1]. Also, PGRs such as auxin or abscisic acid affect ethylene production [112]. However, few attempts have been made to use ethylene directly to stimulate flowering. Ethrel (2-chloroethylphosphonic acid) at 200 mg/l has been applied to *Cupressus arizonica* Greene and *Chamaecyparis lawsoniana* (A. Murr.) Parl. [3]. It strongly enhanced the effectiveness of GA_3 in promoting both male and female cone production but was without effect when applied alone. But when Ethrel was applied to *Thuja plicata* Down ex D. Don at an early stage of initiation, it caused female buds, orginally induced by GA_3, to revert back to vegetative buds [47]. A synergistic effect of Ethrel on GA_3 induction was observed on *Cryptomeria japonica* D. Don. [30] and, to some extent, on loblolly pine [26]. Remrod [85] doubled the number of female cones produced on grafts of Norway spruce by treatment with Ethrel alone (200mg/l spray) but there was no effect on production of male cones. No stimulating effect of Ethrel was found on Douglas-fir, but the timing of the treatment relative to the application of GAs alone was not studied [6]. Timing can be an important factor as mentioned above for the use of growth retardants. Hashizume [30] suggested that the effect of Ethrel should vary with tree age and should be effective mainly in young seedlings.

Cytokinins have been tested in a few cases. In Douglas-fir [94] and in Sitka spruce [103], N6-benzyladenine enhanced GA-induced flowering. A possible physiological role for cytokinins in flower induction has been discussed by Zaerr and Bonnet-Masimbert [113].

It is difficult to predict whether or not PRGs other than NAA and GAs will be useful in stimulating flowering because too few tests have been reported. Moreover, the timing of the treatments with other PGRs has not been satisfactorily tested (see examples with CCC above). In any case, GAs will continue to be the major PGR used to induce flowering.

2.4. Adjunct cultural treatments

Environmental conditions are often unfavorable for flowering. Under such conditions (e.g., cold and cloudy weather at the time of bud initiation and differentiation), and with very young plants, it may be necessary to apply adjunct treatments which are synergistic with GAs. These treatments are often those that were applied with moderate success, before the recent success with GAs. We discuss here only reports dealing with both GAs and cultural practices.

Branch or trunk bark ringing (girdling), water stress, and root pruning are treatments which are commonly used with great efficacy on many species such as Douglas-fir [69, 87], lodgepole pine [111], loblolly pine [27], western hemlock [97], and Sitka spruce [46]. It is important to note that girdling may produce a carry-over stimulation of flowering into the second year, or even longer [23]. This is also true with GAs + girdling treatments on Douglas-fir [72], larch sp. [4], Sitka spruce [79], and Norway spruce [7]. Root pruning may be very useful for flowering stimulation in seed orchards [98, 109]. Also, root flooding has been effective on Douglas-fir [5].

Environmental modifications, mainly temperature, but also photoperiod and light intensity, are powerful adjunct treatments. They greatly influence the response to hormonal treatments and may play a key role in sex determination (See Ross and Pharis, in this volume). Nagao [57, 58, 59] applied different combinations of day and night temperatures under lights of different color (with a constant light energy) to *Cryptomeria* and *Chamaecyparis* trees that were also treated with GA_3 (and occasionally with $GA_{4/7}$). He demonstrated a clear effect of both temperature and light conditions on the response to the GA treatment for both vegetative growth and reproductive development. These treatments differ greatly in sex determination. Pollard and Portlock [81], treated young seedlings and cuttings of western hemlock with $GA_{4/7}$ and demonstrated the effect of photoperiod and temperature on intensity of flowering and sex determination. Interestingly, the temperature and photoperiod after the induction treatment strongly influenced the flowering response. This observation can explain the differences encountered by physiologists in their attempts to correlate flowering and climatic factors. Kosinski and Giertych [35], using optical fibers, have clearly demonstrated the effects of light intensity on the flowering of Scots pine and Norway spruce, and Bonnet-Masimbert (unpublished results, 1986) found that increased light intensity strongly stimulates both male and female flowering of Douglas-fir.

Practical treatments for increased flowering include placing in a greenhouse (probably through increased temperature), accelerated growth regimes, and nitrate fertilization on Norway spruce [15], Sitka spruce [104], western hemlock [80], and jack pine [10]. Out-of-phase dormancy treatment was proposed by Greenwood [27] for loblolly pine. He prolonged shoot elongation during winter by applying a 20-hr photoperiod and increasing the temperature, then induced the formation of resting buds in February by shortening the photoperiod. He was thus able to stimulate both male and female flowering compared to only female stimulation with $GA_{4/7}$ plus water stress. Provided it is properly timed, high temperature is a very powerful treatment in conjunction (but often not at the same time) with water stress (e.g., [90] on mature grafts of Engelmann spruce (*Picea engelmannii* Perry ex Engelm.), treated with $GA_{4/7}$).

The timing of the adjunct treatments strongly influences their effectiveness on loblolly pine [28], white spruce [53], and Engelmann spruce [90]. It is clear that a better understanding of the physiological effect of these kinds of treatments is urgently needed in order to make better use of them.

2.5. Relation with the growth and the general development of the plants

It is clear that under natural conditions the potentials of bud primordia for sexualization vary with location in the crown. It is therefore important to take into account the position in the crown when interpreting flowering experiments (larch [4]; different species of spruces [7, 52, 79, 90]). It can also be important to apply flower-inducing treatments to the specific parts of the plant that have a propensity to flower, (Ross, personal communication, 1986) [37, 40]. However, after treatment with GAs it is not rare to observe flowers on those parts of the trees where they usually do not appear (e.g, female cones in the lower part of larch [4] and western hemlock [97] trees).

A more important consideration is the time of year when bud differentiation occurs. Determination of the time when the first male and female parts are detectable is very helpful, but this information alone cannot indicate when the flower differentiation process commenced [61]. Recently Owens *et al.* [62, 63] have shown that one effect of a flower inducing treatment (root pruning or root pruning + $GA_{4/7}$) was to delay the development of axillary apices. Observing this modification of the normal course of development might also help determine the proper timing of treatments, but such observations would be difficult. Another complicating aspect of timing is that, depending on the kind of treatment applied, the optimal periods may change. A good example of this kind of complication is reported by Ross [90] on Engelmann spruce; it appears that under $GA_{4/7}$ treatment the critical timing for formation of flowers by high temperature (i.e., late stage of slow shoot elongation) is not the same as when flowering is promoted by drought (i.e., early and rapid shoot growth). There is also an interaction with temperature conditions [77]. It might be possible to link the timing of treatment to developmental criteria, such as individual growth curves [53, 90], number of needles, or size of buds [43]. In addition, it seems that all the parts of the crown of a given tree are not synchronized with regard to their "critical period" for flower-inducing treatments. It is possible that the period of flower bud formation is longer than previously thought and that not all the sexual buds are initiated at the same time (Ross, personal communication, 1986).

There seems to exist a relationship between shoot elongation and response to treatment with GAs [37, 53, 89, 95]. It has been postulated that GAs are preferentially used by conifers for vegetative growth practices. Only when environmental conditions, or other factors such as maturation and aging, restrict vegetative growth may endogenous GAs become available for a long enough period and in sufficiently high concentrations for cone bud differentiation to occur [66, 67, 70, 95]. This hypothesis is in agreement with observations on clones of Douglas-fir which had a poor or a good history of flowering [109]. The growth of "good flowering" trees is not stimulated by external application of $GA_{4/7}$ as opposed to "poor flowering" trees, which exclusively respond to GA treatment by increasing their shoot growth. Another interesting factor may be, at least for the Pinaceae, the environmental conditions of the year prior to treatment and their effect on growth and bud development. In an experiment on 5-year-old seedlings of Douglas-fir which had not flowered previously, Lallier [37] applied $GA_{4/7}$ in May, 1981. Trees that subsequently flowered in 1982 had a reduced shoot elongation the year of treatment (1981) compared to the nonflowering trees, showing the classical antagonism between growth and flowering (Figure 1). But interestingly, the situation was inverse the previous year. With respect to nonflowering trees, those which flowered in 1982 had reduced shoot growth the year of treatment (1981) but greater shoot growth the previous year (1980). Whether this effect was due to an increased size of buds, and through this an increased number of potential axillary primordia for the next year, or whether it had only to do with trophic problems is still under study (Bonnet-Masimbert, unpublished data, 1985). This observation suggests that specific pretreatment of trees the previous year may predispose them to respond to flower-inducing treatments.

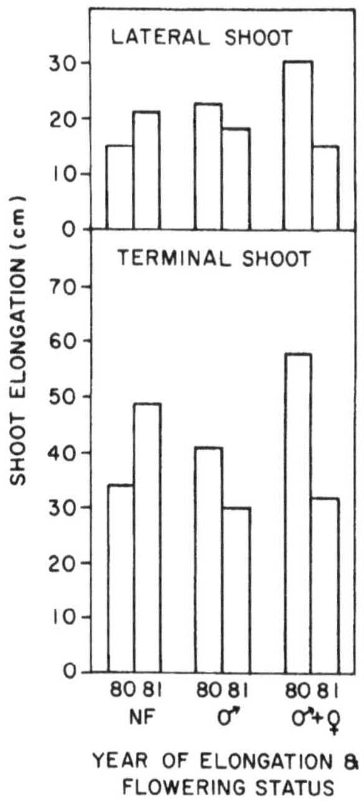

Figure 1. Shoot elongation of nonflowering trees (NF) and flowering trees with only male strobili (♂) or both male and female strobili (♂ + ♀) in 1980 and 1981. Flowering was induced by a mixture of $GA_{4/7}$ + NAA in the spring, 1981, and observed in spring 1982 [37].

Finally, the relationship between root activity and a tree's ability to respond to hormonal treatment may provide a clue to future flower-inducing techniques. As shown above, many of the adjunct treatments commonly applied can be interpreted as affecting root growth. Whether this relationship is direct [5, 77] or indirect [95] is still unclear. On Douglas-fir, the classical bark girdling and root pruning treatments stop root growth [37], as does flooding [5]. Recently, treatment with $GA_{4/7}$ was observed to block root growth (Figure 2) (Bonnet-Masimbert, unpublished results). Further hormonal analyses, in conjunction with precise measurements of shoot and root growth [5], might clarify this point.

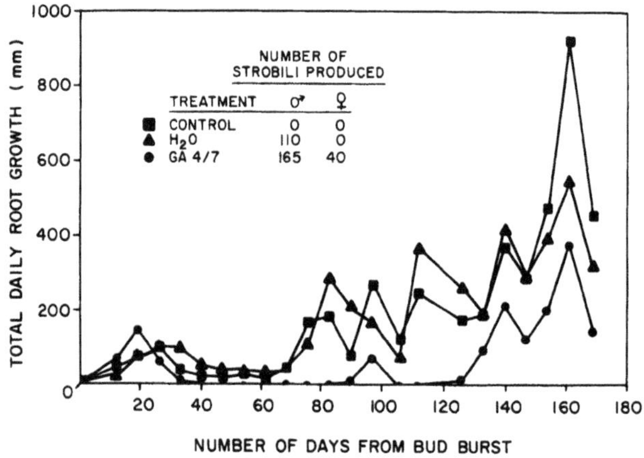

Figure 2. The effect of different treatments on root growth and flowering response of three ramets of one clone (# 1080) of Douglas-fir. The root growth was observed in individual rhizotrons [37], and is expressed as the number of actively growing roots multiplied by the mean daily increase in length of each root (calculated from observations on 20 roots). The treatments applied at bud burst were: control (□), stem injection of 100 ml of a 5% ethanol solution in water (△), stem injection of 100 mg of $GA_{4/7}$ in 100 ml of a 5% ethanol solution in water (○) (Bonnet-Masimbert, unpublished results, 1984).

3. ENDOGENOUS GROWTH REGULATORS AND FLOWERING

It is now clear that PGRs, and especially GAs, constitute one of the possible keys to manipulating flowering, particularly in conifers [70]. But whether or not changes in endogenous PGRs accompany flower initiation is not well documented. Although we know something about the use of PGRs to induce flowering, we know little about their physiological role. Successful exogenous application of GAs does not necessarily indicate their endogenous role.

Kulikowska *et al.* [36], and Ivonis *et al.* [31] found that both GAs and auxin may modify the time of cone bud differentiation in Scots pine and Norway spruce. It has also been demonstrated that flowering induction methods such as nitrate fertilization or water stress in Douglas-fir [24, 70] or the use of a polyethylene cover on Norway spruce [16, 22] are accompanied by an increase in GAs that are less polar than GA_3. These kinds of results have led to improved methods for stimulating flowering in a number of Pinaceae species.

3.1. Flowering and changes in endogenous GAs

Most physiological analyses reported to date have focused upon GAs. Whether or not the applied doses of GAs are much higher than endogenous levels [55, 102], implying a possible pharmacological effect instead of a physiological one, has been recently discussed by Ross and Pharis [95]. It seems that, due to a low rate of absorption by the treated trees and rapid metabolism or conversion to inactive forms of GAs, only a very small part of the applied GAs reach differentiating meristems [95]. Hypotheses concerning penetration or conversion have been confirmed by the use of tritiated GAs [22, 70, 107]. But unless very precise quantitation of GAs in meristem tissues is accomplished, this question cannot be answered satisfactorily.

Puritch *et al.* [83] have suggested that vital metabolites may be used to develop the reproductive buds at the expense of vegetative buds or that hormone applications may affect the concentrations of other growth substances. It has also been proposed by Lee *et al.* [38] that the floral stimulation in Korean pine and Japanese larch by GA treatment may be mediated through an increased ratio of soluble sugars to amino acids, which is caused by reduced bud activity at the time of floral initiation. Interestingly, Ross *et al.* [91] have shown in radiata pine that one effect of treatment with GAs was a reallocation of photoassimilates in favor of long-shoot primordia (the potential cone buds) at the expense of the apical dome. But these authors also demonstrated that, as opposed to GA_3, $GA_{4/7}$ had not only the reallocation effect but also a specific morphogenetic function in sexual differentiation. It would appear that these two effects are independent. In any event, experiments at the level of the meristems, in association with precise anatomical studies (e.g., [62, 63]), and, if possible, immunolocalization of growth substances [101, 114], are necessary for a better understanding of the delicate relationship between vegetative growth and reproductive developmental processes.

In most cases, GAs do not seem to be specifically used by plants to control their flowering. Many experiments have suggested that GAs could even be preferentially used for vegetative growth processes [65, 70]. This preference seems to decrease as the tree matures [89]. But chronological changes in the GA levels may be more important than the actual concentrations at any given time. Phenological observations associated with the flower-inducing treatments need to be examined in terms of changes in the concentrations in GAs and other PGRs. Such studies are rarely reported.

Chalupka *et al.* [16] analyzed the effect of a polyethylene cover on Norway spruce grafts, and showed a rapid (one day after covering) increase of all the GAs. This effect persisted for at least 2 weeks, but only for the less polar GAs. On the same species, Dunberg *et al.* [22], using tritiated GA_4 and GA_9, showed that one effect of the plastic covering in the greenhouse was a retarded metabolism of GA_4. They hypothesized that flowering may be stimulated by this kind of metabolism.

Wesoly [110] demonstrated that branch girdling of Scots pine, a treatment which stimulates flowering, is accompanied within 48 hours by an increase of the less polar GAs. After 48 hours, the level of these GAs was equal to or greater than the levels observed only 2 weeks later in the ungirdled branches. The content in endogenous auxin was also analyzed. No significant effect of girdling was observed, but there were large clonal variations. In these studies, all quantitation was accomplished with bioassays, which are useful for comparing PGR levels in similar tissues. It is safer, however, to confirm bioassay results with more rigorous chemical or immunological techniques. Despite the shortcomings of bioassays, these results are evidence that the metabolism of GAs is influenced by treatments that influence flowering. Pharis *et al.* [76] recently reported increased levels of less polar endogenous GAs for several cone enhancing treatments such as drought, girdling, N fertilization, and root pruning.

3.2. Flowering and changes in different groups of PGRs

The hypothesis that PGRs besides the GAs may be involved in control of the flowering process, and especially the possibility that changes in the balance between these regulators may be important, has led to a number of recent reports.

Galoch [25] studied the juvenility in birch in terms of changes in GAs, ABA, IAA, and cytokinins (CKs), using chromatographic techniques and bioassays. Comparing 4-year-old vegetative plants, flowering plants, and 15-year-old mature trees, she found drastic variations in the content of PGRs in buds and apical sections of shoots. Especially striking was an increase of ABA in young flowering trees. Young nonflowering trees had a high level of GAs in apical buds and a very low level in axillary buds. In contrast, young flowering trees and mature ones had very reduced differences between these two types of buds, due to an increase of gibberellins in axillary buds, as opposed to a slight reduction of these regulators in apical buds. There was little difference in auxin concentrations. Also, in this hardwood species, juvenile vegetative plants showed the highest level of CKs, and fully mature 15-year-old plants exhibited the lowest level.

McMullan [55] compared the concentrations of endogenous auxin, ABA, and GA-like substances in buds and elongating shoots (as estimated by bioassays) of poor and good cone-producing Douglas-fir trees. She also applied crude extracts of buds or shoots of good cone-producing clones to young trees in order to test their possible effect on flowering. This last part of the experiment demonstrated only some effect on induction of male flowers. Although there were big differences in cone production by the two clones, most differences in analyzed PGRs between good and poor cone-producing clones were small. Large variations in relation to stage of development were observed in both clones. Although promoters and inhibitors may overlap in chromatograms, which can obscure actual differences, she concluded that differences in endogenous growth substances did not provide an explanation for differences in cone production ability of these two Douglas-fir clones.

A series of analyses has been started recently on Douglas-fir using immunological (RIA and ELISA) methods for quantitations of three classes of PGRs separated by column chromatography and high-performance liquid chromatography (HPLC): ABA, IAA, and CKs (Bonnet-Masimbert, Doumas, Imbault, Joseph, Miginiac, Pilate, Zaerr, unpublished results). The techniques employed were developed by Leroux *et al.* [39], by Maldiney *et al.* [51] for simultaneous quantitation of ABA, auxin and zeatin-riboside (ZR), and by McDonald *et al.* [54] for a detailed study of the different CKs. Very recently, some analyses have also been accomplished using a new anti-GA_3 antibody (Sotta and Pilate, unpublished results). Plant material grown at I.N.R.A., Orleans, France, was, depending on the year of experiment, 4- to 6-year-old grafts or cuttings. When possible, cuttings were used instead of grafted material, especially because some treatments were applied to roots. Use of completely identical genotypes (i.e., the same one for shoot and root) is likely to be important.

As already discussed in this paper, root activity may participate in the reproductive processes of trees. Crozier and Reid [18, 19] reported bioassay evidence that root tips may be involved in interconversion, synthesis, or release of GA-like substances, some of which may be exported to the top. Also, Skene [100] and Torrey [105] have shown that levels of CKs and GAs in shoots are influenced by root environment and activity. It thus appears that root activity may influence flowering. This experiment hypothesized that growth regulators may be involved in this effect [100].

In one trial commenced in 1984, trees of each clone were subjected at the time of bud burst to one of four treatments:

1. Control.

2. Stem injection of 100 ml $GA_{4/7}$ (100 ppm) and NAA (10 ppm) solution.

3. Alternate root flooding (2 1/2 days in water, and 2 1/2 days out) for 3 weeks.

4. Combination of treatments 2 and 3.

These treatments are known to stimulate flowering [5] as confirmed by the flowering response of this 1984 experiment, which is given on Table 1 (Bonnet-Masimbert, unpublished results).

Three and 6 weeks after the beginning of the treatments (bud burst), two elongating shoots in the upper crown of each tree were collected. Each shoot was separated into the distal (upper) one-third part and the proximal (lower) two-thirds. This was done to take into account the natural sexual gradient which is observed along the twig.

Presently, most of the analyses have been made on two ramets of the same clone having received the four treatments, but some analyses on two other clones have also been completed. Thus, only preliminary results can be reported here. It appears that concentrations of all the PGRs analyzed are modified by the treatments. Whether or not concentrations are directly related to flowering has still to be ascertained. Tremendous variations between clones, as already observed by Wesoly [110] for auxin, and also between ramets of one clone, complicate the interpretations of results. But two trends may be of interest:

1. Large changes between PGR levels in the apical and basal portions of shoots may be caused by the root treatments.

Table 1. Flowering of six-year-old Douglas-fir clones in response to gibberellin and/or root flooding. 1984 experiment.

	MALE FLOWERING		FEMALE FLOWERING	
TREATMENT	% flowering plants	flowering code*	% flowering plants	number of cones
1. Control	6.4	0.24	3.2	0.16
2. GA + NAA	36.0	1.48	24.0	2.36
3. Root flooding	48.6	1.73	20.0	3.95
4. Root flooding + GA + NAA	78.1	3.12	43.7	7.49

* For male flowering, the code shown is the mean value of the coded score for each tree in which: 1 = 0 to 10 strobili; 2 = 11 to 30; 3 = 31 to 60; 4 = 61 to 100; 5 = >100.

2. Different treatments cause different responses in PGRs at 3 weeks more than at 6 weeks after treatment.

Control trees have much lower concentrations of ABA than treated trees (from 700 and 400 ng/g dry weight, respectively, for apical and basal portions in control to 3,000 and 5,000 ng/g in treated trees). The increase was especially strong for the basal portions of the shoots (Pilate, unpublished results). We might expect this for flooding, a treatment known to increase ABA [56, 106], but treatment with $GA_{4/7}$ + NAA gave an even stronger response. It is possible that this increase in ABA was a wound response to the method of stem injection. This increase was greater after 6 weeks than after 3 weeks.

In other experiments on Douglas-fir [108], the lowest concentration of ABA occurred just before bud burst. Also, McMullan [55] observed a general increase of ABA toward the end of May and early June, as the rate of shoot elongation decreased. Inductive treatments may have caused an earlier increase in ABA levels which could have brought about decreased shoot growth. Owens *et al.* [63] observed on this same species that one of the effects of root pruning was to retard the development of axillary apices. It appears paradoxical that both GAs and ABA (which is known as an antagonist of the GAs [17]) might positively affect the flowering process.

Among four flowering and four nonflowering trees (Figure 3), there was a higher concentration in total cytokinins after 3 weeks in the apical part of shoots of the flowering trees (Figure 3b). At that time, effects on concentrations of isopentenyladenine and cytokinins in the ZR-group were even more striking (Figure 3a) (Bonnet-Masimbert, Imbault, Joseph, unpublished results). After 6 weeks, the differences were less pronounced

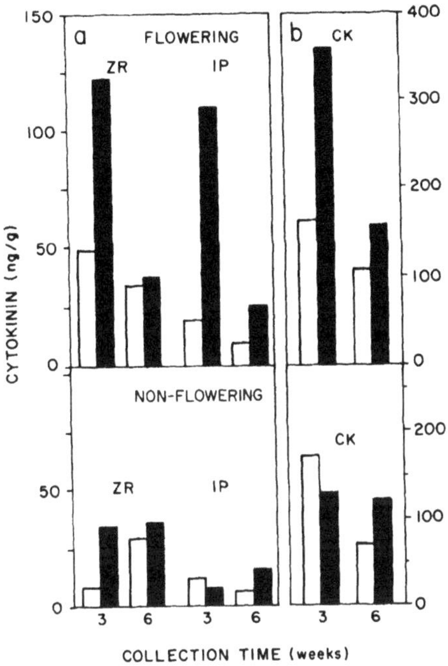

Figure 3. Cytokinin content of shoots from four flowering (upper) and four nonflowering (lower) trees. Basal (□) and apical (■) portion of the shoots at three and six weeks after bud burst.

a - Results for zeatin-riboside (ZR) and isopentenyl adenine (IP).

b - Results for the total cytokinins.

The flowering or nonflowering responses were obtained after treatments numbers 2, 3, 4 and 4 (flowering trees) and numbers 1 & 1, 2, 3 (nonflowering trees) shown in Table 1 (Bonnet-Masimbert, Imbault, Joseph, unpublished results, 1986).

In another experiment [113], cytokinins were analyzed 6 weeks after bud burst in similar shoots. All trees with female flowers had low concentration of total cytokinins, although low CK content did not guarantee flowering. A very striking difference between the basal part of the shoots and the apical part was also demonstrated.

A new cytokinin conjugate, reported by Morris *et al.* (in preparation), which is present in roots, sap, shoots and needles, might play a role in flower differentiation [20]. It is apparently a hexose conjugate of ZR, but the precise structure of the hexose moity has not been elucidated. Such a conjugate could serve as a storage pool for ZR. Interconversion between the conjugate and ZR could be a mechanism by which plants regulate the concentration of ZR. Thus, rapid and reversible changes in levels of ZR could be easily brought about.

Gibberellin analyses using GA_3 antibodies have so far been completed on six trees in the 1984 experiment: one control tree, two GA-treated ones, two flooded ones, and one tree treated with GA + flooding (Pilate, unpublished results). Only the results from samples taken 3 weeks after treatment are available, but it appears that gibberellin activity occurs in a part of the HPLC chromatogram containing GAs less polar than GA_3 (i.e., GA_4). These GA antibodies have a high degree of cross reactivity with GA_4. This activity (concentrations as high as 2000 ng/g) was present only in those three trees that received the $GA_{4/7}$ treatment, with or without flooding. Of those three trees, only two flowered; another tree (flooded without $GA_{4/7}$) flowered profusely but exhibited only low GA activity in this part of the chromatogram. This reduced GA level may be due to a temporarily reduced or delayed production of GAs caused by flooding [84]. Subsequent analyses will be necessary for a better interpretation, but it is clear that one of the flowering trees had low levels of gibberellin at a time when the other two had very high levels. The chemical identification of these GAs is in progress.

These results must be considered as preliminary. More trees belonging to a greater number of clones must be analyzed, and the period of bud differentiation must be sampled more intensively. But these results point up the potential of more precise immunological techniques for the several groups of PGRs, and they suggest some interesting trends. GAs are apparently not the only PGRs to be affected by inductive treatments, and the role of GAs in flower induction remains obscure. It is clear, however, that anatomical analyses should be included in future studies [62, 63].

4. CONCLUSION

Plant growth regulators, particularly the GAs, are becoming useful and more reliable for stimulating flowering of conifers. A strong interaction between some environmental factors (mainly temperature, light and atmospheric humidity) and the response to treatment with PGRs [77] makes these methods especially appropriate for use in a greenhouse setting. The methods are useful to forest tree breeders and they can be used in container seed orchards [92]. However, good results have also been obtained on larger trees in more common outdoor seed orchards [78, 98], generally after the combination of treatment with PGRs and a cultural treatment such as bark girdling or root pruning. Thus, the use of PGRs is possible and practical, for some species. But for many other species, work has been limited to the experimental level and results are very uncertain.

Several important questions need attention if we are to manipulate the flowering process effectively: 1) What are the mechanisms involved in the control of PGRs over flowering processes? 2) Are there PGR's other than GAs taking part in this control? 3) Are there control mechanisms other than those involving the known PGRs? A more cooperative research effort between the physiologists and geneticists interested in the flowering of trees would seem to be a useful step toward answering these questions and contributing to the production of genetically improved forest trees.

REFERENCES

1. Abeles FB (1973) Ethylene in plant biology, 302 p. New York: Academic Press,
2. Bleymüller M (1976) Investigations on the dependence of flowering in spruce [*Picea abies* (L).Karst] upon age and hormone treatment. Silvae Genet 25:83-85
3. Bonnet-Masimbert M (1971) Induction florale précoce chez *Cupressus arizonica* et *Chamaecyparis lawsoniana*. Silvae Genet 20:82-90
4. Bonnet-Masimbert M (1982a) Effect of growth regulators, girdling and mulch on flowering of young European and Japanese larches under field conditions. Can J For Res 12:270-279
5. Bonnet-Masimbert M (1982b) Influence de l'état d'activité des racines sur la floraison induite par des gibbérellines 4 et 7 chez *Pseudotsuga menziesii* (Mirb.) Franco. Silvae Genet 31:178-182
6. Bonnet-Masimbert M (1983) Stimulation de la floraison chez les coniféres: perspectives offertes par l'utilisation des régulateurs de croissance. Coll COLUMA "Les substances de croissance et leurs utilisations en agriculture" Paris, ed. ACTA. Paris, t.2:480-486
7. Bonnet-Masimbert M (1986) Preliminary results on gibberellin induction of flowering of seedlings and cuttings of Norway spruce indicate some carry-over effects. In: For Ecol Manage, IUFRO Symposium on Flowering and Seed Bearing in Forest Seed Orchards, Kornik, Poland, Sept, 1985 (in press)
8. Bower RC and Ross SD (1985) Evaluation of ULV sprays for seed orchard application of $GA_{4/7}$. In: 20th Mtg Can Tree Improvement Assoc Proc, Part 2, 19-22 Aug, Laval, Quebec, Ottawa, Ont. Can Forest Service (in press)
9. Brix M and Portlock FT (1982) Flowering response of western hemlock seedlings to gibberellin and water-stress treatments. Can J For Res 12:76-82
10. Cecich RA (1981) Applied gibberellin $A_{4/7}$ increases ovulate strobili production in accelerated growth Jack pine seedlings. Can J For Res 11:580-585
11. Cecich RA (1983) Flowering in a Jack pine seedling seed orchard increased by spraying with gibberellin $A_{4/7}$. Can J For Res 13:1056-1062
12. Cecich RA (1985) White spruce (*Picea glauca*) flowering in response to spray application of gibberellin $A_{4/7}$. Can J For Res 15:170-173
13. Chalupka W (1979) Effect of growth regulators on flowering of Norway spruce [*Picea abies* (L) Karst] grafts. Silvae Genet 28:125
14. Chalupka W (1978) Effect of growth regulators on the flowering of Scots pine (*Pinus sylvestris* L) grafts. Silvae Genet 27:62-65
15. Chalupka W and Giertych M (1977) The effect of polyethylene covers on the flowering of Norway spruce [*Picea abies* (L) Karst] grafts. Arbor Kornickie 22:185-192
16. Chalupka W, Giertych M and Kopcewicz J (1982) Effect of polyethylene covers, a flower inducing treatment, on the contents of endogenous gibberellin-like substances in grafts of Norway spruce. Physiol Plant 54:79-82
17. Chrispeels MJ and Varner JE (1967) Hormonal control of enzyme synthesis: on mode of action of gibberellic acid and abscisin in aleurone layers of barley. Plant Physiol 42:1008-1016

18. Crozier A and Reid DM (1970) Gibberellin metabolism in the roots of *Phaseolus coccineus* seedlings. In: Carr DJ, ed. Plant Growth Substances, pp. 414-419. Berlin, Heidelberg, New York: Springer Verlag
19. Crozier A and Reid DM (1971) Do roots synthetize gibberellins? Can J Bot 49:967-975
20. Doumas P, Morris JW, Chien C, Bonnet-Masimbert M and Zaerr JB (1986) A possible relationship between a cytokinin conjugate and flowering in Douglas-fir. In: 9th North American Forest Biology Workshop, June 1986, Oklahoma State Univ, Stillwater
21. Dunberg A (1980) Stimulation of flowering in *Picea abies* by gibberellins. Silvae Genet 29:51-53
22. Dunberg A, Malmerg G, Sassa T and Pharis RP (1983) Metabolism of tritiated gibberellin A 4 and A 9 in Norway spruce, *Picea abies* (L) Karst. Effects of a cultural treatment known to enhance flowering. Plant Physiol 71:257-262
23. Ebell LF (1971) Girdling: its effect on carbohydrate status and on reproductive bud and cone development in Douglas-fir. Can J Bot 49:453-466
24. Ebell LF and McMullan EE (1970) Nitrogenous substances associated with differential cone production responses in Douglas-fir to ammonium and nitrate fertilization. Can J Bot 48:2169-2177
25. Galoch E (1985) Comparison of the content of growth regulators in juvenile and adult plants of birch (*Betula verrucosa* Ehrh). Acta Physiologiae Plant 7(4):205-215
26. Greenwood MS (1978) Flower stimulation techniques for loblolly pine (*Pinus taeda* L). In: Proc Third World Consult For Tree Breeding, March, 1977, Vol II, CSIRO, Canberra, Australia, pp. 1031-1042
27. Greenwood MS (1981) Reproductive development in loblolly pine. II. The effect of age, gibberellin plus water stress and out-of-phase dormancy on long shoot growth behavior . Am J Bot 68:1184-1190
28. Greenwood MS (1982) Rate, timing and mode of gibberellin application for female strobilus production by grafted loblolly pine. Can J For Res 12:998-1102
29. Hashizume H (1967) Experimental induction of female flowers in young Japanese larch (*Larix leptolepis* Gordon). Nippon Rin Gakkaishi 49:405-508
30. Hashizume H (1975) Interrelationship of gibberellin and ethrel in the flower bud formation and growth of conifers. Bull Tottori Univ For 8:1-10
31. Ivonis IYu, Kyalina LV, Litvinova VB and Khokhlina EV (1982) Annual dynamics of gibberellin-like substances in needles of Norway spruce clone of different sexes. Soviet Plant Physiol 28:915-920. [transl from Fiziologiya Rastenii 28(1981): 1258-1265]
32. Jackson DI and Sweet GB (1972) Flower initiation in temperate woody plants. Hortic Abstr 42:9-24
33. Kanekawa T and Katsuta M (1982) Promotion of strobilus production in *Pinus thunbergii* Parl and *P densiflora* Sieb et Zucc by gibberellins. J Jap For Soc 64:101-106
34. Kato Y, Miyake I and Ishikawa H (1958) Initiation of flower bud by gibberellin in *Cryptomeria japonica*. J Jap For Soc 40:35-36
35. Kosinski G and Giertych M (1982) Light conditions inside developing buds affect floral induction. Planta 155:93-94

36. Kulikowska HJ, Kopcewicz J, Zatorska Z and Szczesniak T (1978) Auxin and gibberellins in embryonic shoots of Scots pine in relation to flower sex differentiation. Acta Soc Bot Pol 47:403-409
37. Lallier N (1982) Induction florale chez les coniféres: étude des interactions entre activité racinaire, croissance aérienne et initiation florale chez *Pseudotsuga menziesii*. DEA Université d'Orléans. Doc Stat Amélioration INRA, Ardon, 45160, Olivet, France, 41 p.
38. Lee KJ, Park JI, Lee JS and Lee SK (1982) Effects of gibberellin 4/7 on carbohydrate and amino acid metabolism in *Pinus koraiensis* and *Larix leptolepis* during the period of floral bud initiation. Res Rep Inst For Genet Korea 18:54-59
39. Leroux B, Maldiney R, Miginiac E, Sossountzov L and Sotta B (1985) Comparative quantitation of abscisic acid in plant extracts by GLC and ELISA method using the avidin-biotin system. Planta 166:524-529
40. Longman KA (1970) Initiation of flowering on first year cuttings of *Metasequoia glyptostroboides* Hu and Cheng. Nature, Lond 227:299-300
41. Longman KA (1976) Some experimental approaches to the problem of phase change in forest trees. Acta Hortic 56:81-90
42. Longman KA (1978) Control of flowering for forest tree improvement and seed production. Sci Hortic 30:1-10
43. Longman KA (1983) Effects of gibberellin, clone and environment on cone initiation, shoot growth and branching in *Pinus contorta*. Ann Bot 50:247-257
44. Longman KA (1985) Variability in flower initiation in forest trees. In: Cannel MGR and Jackson JE, eds. Trees as Crop Plants. Inst Terrestrial Ecol, NERC pp. 398-408
45. Longman KA, Dick J and Page CN (1982) Cone induction with gibberellin for Taxonomic studies in Cupressaceae and Taxodiaceae. Biol Plant 24:3,195-201
46. Longman KA, Dick JMcP, Mugglestone M and Smith RJ (1986) Effect of gibberellin A_{4+7} and bark-ringing on cone initiation in mature *Picea sitchensis* grafts. Tree Physiology 1:101-113
47. Longman KA and Manurung R (1982) Manipulation of cone formation in the Cupressaceae. Ann Rep Inst Terr Ecol 1981:67-69
48. Luukkanen O (1979) Hormonal treatment increases flowering in Norway spruce grafts grown in a plastic greenhouse. Finnish Found For Tree Breeding Ann Rep, pp. 20-26
49. Luukkanen O (1981) Effect of gibberellins A_4 and A_7 on flowering in Scots pine grafts. Silvae Fenn 15:359-365
50. Luukkanen O and Johansson S (1980) Effect of exogenous gibberellins on flowering in *Pinus sylvestris* L grafts. Physiol Plant 50:365-370
51. Maldiney R, Leroux B, Sabbagh I, Sotta B, Sossountzov L and Miginiac E (1985) Enzyme immunoassay for the quantitation of three phytohormones: auxin, abscisic acid and zeatin-riboside using a avidin biotin system. Jour of Immunological Methods (in press)
52. Marquard RD and Hanover JW (1984a) Sexual zonation in the crown of *Picea glauca* and flowering response to exogenous $GA_{4/7}$. Can J For Res 14:27-30
53. Marquard R and Hanover J (1984b) Relationship between gibberellin $A_{4/7}$ concentration, time of treatment and crown position on flowering of *Picea glauca*. Can J For Res 14:547-553

54. McDonald EMS and Morris RO (1985) Isolation of cytokinins by immunoaffinity chromatography and analysis by high performance liquid chromatography-radioimmunoassay. Method in Enzymol 110:347-358
55. McMullan EE (1980) Effect of applied growth regulators on cone production in Douglas-fir, and relation of endogenous growth regulators in cone production capacity. Can J For Res 10:405-422
56. Menyailo LN and Shul'gina GC (1978) Effect of swamp drainage on hormonal metabolism of scots pine. Sov J Plant Physiol 25:1, 94-98
57. Nagao A (1981) Effects of spectral compositions on flower bud formation in *Cryptomeria japonica* under different temperature conditions. J Jap For Soc 63(10):367-371
58. Nagao A (1983a) Effects of different light spectral composition on floral initiation in *Chamaecyparis obtusa* seedlings under various temperature conditions. J Jap For Soc 65(7):233-236
59. Nagao A (1983b) Differences of flower initiation of *Cryptomeria japonica* under various alternating temperatures. J Jap For Soc 65(9):335-338
60. Owens JN and Molder M (1977) Cone induction in yellow cypress (*Chamaecyparis nootkatensis*) by gibberellin A_3 and the subsequent development of seeds within the induced cones. Can J For Res 7:605-613
61. Owens JN and Molder M (1979) The times and patterns of cone differentiation in North American conifers. In: Bonner FT and Starkville MS, eds. Proc Symp Flowering and Seed Development in Trees. USDA Forest Service, Southern For Exp Stn, pp. 25-32
62. Owens JN, Webber JE, Ross SD and Pharis RP (1985) Interaction between gibberellin $A_{4/7}$ and root pruning on the reproductive and vegetative processes in Douglas-fir. III. Effect on anatomy of shoot elongation and terminal bud development. Can J For Res 15:2, 354-364
63. Owens JN, Webber JE, Ross SD and Pharis RP (1986) Interaction between gibberellin $A_{4/7}$ and root pruning on the reproductive and vegetative processes in Douglas-fir. IV. Effect on lateral bud development. Can J For Res 16:2, 211-221
64. Pharis RP (1975) Promotion of flowering in conifers by gibberellins. For Chron 51:244-248
65. Pharis RP (1977) Interaction of native and endogenous plant hormones in the flowering of woody plants. In: Schutte HR and Gross D, eds. Proc. Conf. Regulation of Developmental Processes in Plants. Halle, Academy of Sciences pp. 343-360.
66. Pharis RP, King RW (1985) Gibberellin and reproductive development in seed plants. Ann Rev Plant Physiol 36:517-568
67. Pharis RP and Kuo CG (1977) Physiology of gibberellins in conifers. Can J For Res 7:299-325
68. Pharis RP and Morf W (1967) Experiments on the precocious flowering of western red cedar and four species of *Cupressus* with gibberellin A_3 and $A_{4/7}$ mixture. Can J Bot 65:1519-1524
69. Pharis RP and Ross SD (1976) Gibberellins: their potential uses in forestry. Outlook on Agric 9:82-87
70. Pharis RP and Ross SD (1986) Hormonal promotion of flowering in the Pinaceae family conifers. In: Halevy A, ed. Handbook of flowering, vol 5. Boca Raton, FL: CRC Press (in press)

71. Pharis RP and Ross SD (1986) Flowering of pinaceae family conifers with gibberellin $A_{4/7}$ mixture: how to accomplish it, mechanisms and integration with early progeny testing. In: Proc Conifer Tree Seed in the Inland Mountain West Symposium, Aug 5-6, 1985, Missoula, Mont. USDA/USFS Intermountain Res Sta. Ogden, Utah 84401. Gen Tech Rpt INT-203, April, 1986 pp. 171-179
72. Pharis RP, Ross SD and McMullan EE (1980) Promotion of flowering in the Pinaceae by gibberellins. III. Seedlings of Douglas-fir. Physiol Plant 50:119-126
73. Pharis RP, Ruddat M, Phillips CC and Heftmann E (1965) Precocious flowering of Arizona cypress with gibberellin. Can J Bot 43:923-927
74. Pharis RP, Tomchuk D, Beail FD, Rauther RM and Kiss G (1986) Promotion of flowering in white spruce (*Picea glauca*) by gibberellin $A_{4/7}$ alone, and with auxin (naphthaleneacetic acid), girdling, and $Ca(NO_3)_2$ as adjunct cultural treatments. Can J For Res 16:340-345
75. Pharis RP, Wample RL and Kamienska A (1975) Growth, development and sexual differentiation in *Pinus* with emphasis on the role of the plant hormone, gibberellin. In: Baumgartner DM, ed. Management of Lodgepole Pine Ecosystems, pp. 106-134. Pullman, WA
76. Pharis RP, Webber JE and Ross SD (1986) The promotion of flowering in forest trees by gibberellin $A_{4/7}$ and cultural treatments: a review of the possible mechanisms. For Ecol Manage, IUFRO Symposium on Flowering and Seed Bearing in Forest Seed Orchards, Kornik, Poland, Sept, 1985 (in press)
77. Philipson JJ (1983) The role of gibberellin $A_{4/7}$, heat and drought in the induction of flowering in Sitka spruce. J Exp Bot 34:291-302
78. Philipson JJ (1985a) The promotion of flowering in large field-grown Sitka spruce by girdling and stem injections of gibberellin $A_{4/7}$. Can J For Res 15:166-170
79. Philipson JJ (1985b) The effect of top pruning, girdling and gibberellin $A_{4/7}$ application on the production and distribution of pollen and seed cones in Sitka spruce. Can J For Res (submitted)
80. Pollard DFW and Portlock FT (1981) Effects of temperature on strobilus production in gibberellin-treated seedlings of western hemlock. Can For Serv Res Notes 1:21-22
81. Pollard DPW and Portlock FT (1984) The effect of photoperiod and temperature on gibberellin $A_{4/7}$ induced strobili production of Western Hemlock. Can J For Res 14:291-294
82. Puritch G (1972) Cone production in conifers. A review of the literature and evaluation of research needs (with an economic analysis by AH Vyse). In Pacific For Res Centre/Can For Service Infor Rep BC-X.65, 94 pp. Victoria, BC
83. Puritch GA, McMullan EE, Meagher MD and Simmons CS (1979) Hormonal enhancement of cone production in Douglas-fir grafts and seedlings. Can J For Res 9:193-200
84. Reid DM, Crozier A and Harvey BRM (1969) The effect of flooding on the export of gibberellins from the root to the shoot. Planta 89:376-379
85. Remrod J (1976) An experiment on flower-induction with Ethrel-Breeding Norway Spruce. Bogesund, pp. 203-205
86. Ross SD (1976) Differential flowering responses by young Douglas-fir grafts and equisized seedlings to gibberellins and auxin. Acta Hort 56:163-167

87. Ross SD (1978) Influences of gibberellins and cultural practices on early flowering of Douglas-fir seedlings and grafts. In: Proc Third World Consult. For. Tree Breeding, March 1977, Vol 2, CSIRO, Canberra, Australia, pp. 997-1107
88. Ross SD (1979) Evaluation of foliar spray formulations for seed orchard application of gibberellins. Weyerhaeuser Forestry Res Tech Rep 042-3001/79/10, pp. 1-12
89. Ross SD (1983) Enhancement of shoot elongation in Douglas-fir by gibberellin $A_{4/7}$ and its relation to the hormonal promotion of flowering. Can J For Res 13:986-994
90. Ross SD (1985) Promotion of flowering in potted *Picea engelmannii* (Perry) grafts: effect of heat, drought, gibberellin $A_{4/7}$ and their timing. Can J For Res 15:4,618-624
91. Ross SD, Bollmann MP, Pharis RP and Sweet GB (1984) Gibberellin $A_{4/7}$ and the promotion of flowering in *Pinus radiata*. Effect on partitioning of photoassimilate within the bud during primordia differentiation. Plant Physiol 76:326-330
92. Ross SD, Eastham AM and Bower RC (1985) Potential for container seed orchards. In: Shearer RS, ed. Proc - Conifer Tree Seed in the Inland Mtn. West Symposium, Missoula, Montana, pp. 180-186. USDA For Serv
93. Ross SD and Greenwood MS (1979) Promotion of flowering in the Pinaceae by gibberellins. II. Grafts of sexually mature and immature *Pinus taeda* L. Physiol Plant 45:207-210
94. Ross SD and Pharis RP (1976) Promotion of flowering in the Pinaceae by gibberellins. I. Sexually mature, non-flowering grafts of Douglas-fir. Physiol Plant 36:182-186
95. Ross SD and Pharis RP (1985) Promotion of flowering in tree crops: different mechanisms and techniques, with special reference to conifers. In: Cannel MGR and Jackson JE, ed. Trees as crop plants. Inst Terrestrial Ecol, NERC, pp. 383-397
96. Ross SD, Pharis RP and Heaman JC (1980) Promotion of cone and seed production in grafted and seedling Douglas-fir seed orchards by application of gibberellin $A_{4/7}$ mixture. Can J For Res 10:464-469
97. Ross SD, Piesch RF and Portlock FI (1980) Promotion of cone and seed production in rooted ramets and seedlings of western hemlock by gibberellins and adjunct cultural treatments. Can J For Res 11:90-98
98. Ross SD, Webber JE, Pharis RP and Owens JN (1985) Interaction between gibberellin $A_{4/7}$ and root pruning on the reproductive and vegetative process in Douglas-fir. I. Effects on flowering. Can J For Res 15:341-347
99. Sato K (1963) Some physiological actions of gibberellins on forest trees. FAO World Consultation For Genet Stockholm. No FAO/FORGEN 63/5-2:1-11
100. Skene KGM (1975) Cytokinin production by roots as a factor in the control of plant growth. In: Torrey JG and Clarkson DT, eds. The development and function of roots, pp. 365-396. London, New York: Academic Press
101. Sotta B, Sossountzov L, Maldiney R, Sabbagh I, Tachon P and Miginiac E (1985) Abscisic acid localization by light microscopic immunohistochemistry in *Chenopodium polyspermum* L. J Histochem Cytochem 33:201-208
102. Sweet GB (1979) A physiological study of seed cone production in *Pinus radiata*. NZ J For Sci 9:20-33

103. Tompsett PB (1977) Studies of growth and flowering in *Picea sitchensis* (Bong) Carr. I. Effect of growth regulator applications to mature scions on seedling rootstocks. Ann Bot 41:1171-1178
104. Tompsett PB and Fletcher AM (1977) Increased flowering of Sitka spruce [*Picea sitchensis* (Bong) Carr] in a polythene house. Silvae Genet 26:84-86
105. Torrey JG (1976) Root hormones and plant growth. Annu Rev Plant Physiol 27:435-459
106. Walton DC (1980) Biochemistry and physiology of abscisic acid. Ann Rev Plant Physiol 31:453-489
107. Wample RL, Durley RC and Pharis RP (1975) Metabolism of gibberellin A_4 by vegetative shoots of Douglas-fir at three stages of ontogeny. Physiol Plant 35:273-278
108. Webber JE, Laver ML, Zaerr JB and Lavender DP (1979) Seasonal variation of abscisic acid in dormant shoots of Douglas-fir. Can J Bot 57:534-538
109. Webber JE, Ross SD, Pharis RP and Owens JN (1985) Interaction between gibberellin $A_{4/7}$ and root pruning on the reproductive and vegetative process in Douglas-fir. II. Effects on shoot growth. Can J For Res 15:348-353
110. Wesoly W (1985) Effect of girdling on flowering and on levels of endogenous growth regulators in embryonic shoots of scots pine grafts (*Pinus silvestris* L). Acta Physiologiae Plantarum 7:4, 171-179
111. Wheeler NC, Wample RL and Pharis RP (1980) Promotion of flowering in the Pinaceae by gibberellins. IV. Seedlings and sexually mature grafts of lodgepole pine. Physiol Plant 50:340-346
112. Yoshii H and Imaseki H (1981) Biosynthesis of auxin induced ethylene: effects of indol-3-acetic acid, benzyl-adenine and abscisic acid on endogenous levels of 1-aminocyclopropane-1-carboxylic acid (ACC) and ACC synthase. Plant Cell Physiol 22:369-379
113. Zaerr JB and Bonnet-Masimbert M (1986) Cytokinin level and flowering in Douglas-fir. In: For Ecol Manage, IUFRO Symposium on Flowering and Seed Bearing in Forest Seed Orchards, Kornik, Poland, Sept, 1986 (in press)
114. Zavala ME and Brandon DL (1983) Localization of a phytohormone using immunocytochemistry. J Cell Biol 97:1235-1239

Plant Growth Regulation 6: 37–60

3. CONTROL OF SEX EXPRESSION IN CONIFERS

S.D. ROSS AND R.P. PHARIS
Senior Scientist, Research Branch, British Columbia Ministry of Forests and Lands, 1320 Glyn Road, Victoria, B.C. V8Z 3A6 and Professor of Botany Physiology, Department of Biology, University of Calgary, Alberta, Canada T2N 1N4

ABSTRACT
Despite considerable progress in recent years in the promotion of flowering in conifers, relatively little is known about the endogenous control of sex expression or its practical manipulation to aid pollen and crop management in seed orchards. Well defined patterns of sexual zonation exist within the tree crown and shoot, and these appear to be associated with hormonal and possibly nutritional gradients. No firm conclusions, however, are possible regarding the specific role(s) of different PGRs in sex expression, this being as much as a problem of lack of critical study as the complexity of the process itself. Whether a PGR such as gibberellins or auxins, or a cultural treatment that itself may influence endogenous hormone levels, primarily promotes female or male flowering often depends on the position within the crown of the shoot being treated, and its timing in relation to stage of primordia development. Photoperiod also may be involved-- short days favoring females and long days males, at least in some species-- although its role in controlling sex expression is by no means clear. Practical manipulation of sex expression may be possible through selective crown pruning, different PGR combinations, treatment timing and/or photoperiod control, but the development of such techniques for operational use will require a better basic understanding of the endogenous regulatory mechanisms involved.

1. INTRODUCTION

The control, hormonal and otherwise, of sex expression in conifers, and especially those commercially important species of the Pinaceae, has received little attention. This is no doubt due to the general, and heretofore more pressing problems, of overcoming a delayed and inadequate flowering. However, promotion of early and enhanced flowering is now possible for many, if not most conifers through the properly timed application of specific gibberellins (GAs), usually administered with adjunct cultural treatments (e.g. nitrogen fertilization, root-pruning, girdling, drought and heat) [see Chapter 2 by Bonnet-Masimbert and Zaerr; also reviews by Owens and Blake (85) and Ross and Pharis (115,116)]. Except in very young seed orchards, where pollen production naturally is sparse, the proper species-dependent $GA_{4/7}$ + cultural treatment will generally provide a good balance of female and male strobili.

However, with the growing interest in artificial pollination, both to improve seed yields and to maximize genetic gains through control of pollen parentage, an ability to regulate the sex of strobili produced becomes increasingly important. Sweet and Krugman (131) discussed the advantages of maintaining separate seed and pollen orchards, with the latter being situated in warmer climates to ensure that pollen matures earlier, thereby allowing time for its collection and processing before the females become receptive. The ability to specifically promote female or male strobili on different individuals will be even more valuable in containerized seed orchards, which presently are being evaluated as a cost-effective alternative to conventional field orchards (9,117).

The literature on sex expression in conifers is presently confused and yields few firm conclusions about its regulation by the plant per se or by environmental factors. Nor is its possible practical control through plant growth regulator (PGR) application or other means clearly discussed in the literature. In this chapter we review the spatial and temporal patterns of sex expression common to representative conifer species, and consider their possible hormonal, nutritional and environmental bases. We also identify promising areas for future research.

2. PATTERNS OF SEX EXPRESSION

Nearly all conifers are predominantly monoecious, although within any species are to be found certain genotypes that produce only female or male strobili [see Owens and Hardev (86)]. And, even within monoecious individuals, there exist spatial and temporal patterns of sex expression that transcend most species and family boundaries (86). A better knowledge of these patterns can not only provide insight into the physiological basis of sex expression, but may also give us the practical means for its control. On the other hand, failure to take these general patterns into account when sampling or designing experimental treatments has been a confounding factor in many studies.

2.1. Spatial and Ontogenetical Patterns

It is true of most conifers that, following a so-called "juvenile" phase of strictly vegetative development lasting several years to several decades, depending on the species [see discussion in Zimmerman *et al.* (147)], regular female flowering preceeds regular male flowering by several years (32,33,140,147). Occasionally, one finds examples of extremely precocious male flowering at ages of one year or less from seed. Although this appears to be an accidental phenomenon in response to some unique inductive condition, a better understanding of its control could be very useful in designing manipulative treatments. Such precocious male-flowering individuals will usually revert to the non-flowering condition and the normal sequence of regular female then male flowering (33).

As with all generalizations there are exceptions. For example, in the Pinaceae regular male flowering typically occurs first in *Pinus radiata* (G.B. Sweet, per. comm.), and we have observed a similar tendency for many Cupressaceae and Taxodiaceae conifers.

In an interesting experiment Giertych (32,33) compared the development on seedling rootstock of grafted scions representing 'strong' and 'weak' shoots (with regard to vegetative vigor) from, respectively, the female and male crown zones of a mature *Pinus sylvestris* tree. Contrary to expectation, the "male" scion did not begin producing male strobili before female strobili. Nor did they come into male flowering at an earlier age than the "female" scion. Both scion types exhibited the typical pattern of regular female followed by male flowering, but the transition occurred more rapidly in the female scion concomitant with their also more vigorous vegetative development. Giertych (34) concluded that factors other than ontogenetic ageing were thus responsible for the within-crown pattern of sex expression, and principal among them was the shoot's vegetative vigor.

Development of patterns of sexual zonation within the tree crown is similar for most conifers (61,70,118,120,140). As the tree matures and increases in size, female strobili are initiated/differentiated first on strong leading shoot in the upper crown. In some species (e.g. Pinus and Tsuga), this may include the main-stem terminal, whereas in others (e.g. Picea and Pseudotsuga) female strobili are first produced on somewhat less vigorous lateral shoots of the current-year leader, and on the terminal shoots of subtending first-order branches. With increasing age of the tree, the zone of female flowering extends downward and initially inward to include still vigorous shoots at the base of the branch. However, with further increasing branch age and declining vigor of some higher-order shoots, the zone of female flowering again retreats outward to include usually only the most vigorous distal shoots. Female flowering, in fact, usually terminates in the lower third of the crown in older trees. Development of the male zone follows the reverse pattern, male strobili being produced first on less vigorous shoots in the distal region of lower branches. But, as the tree ages their production extends upward and inward, usually terminating in the middle crown. There is commonly a transition zone where both types of strobili occur on the same branch, and in some species even on the same internodal shoot, but with females always in the more distal position. Again there are exceptions. Nagao et al. (81) note the existence of some unusual Chamaecyparis obtusa clones which bear female strobili in the lower crown and male strobili in the upper crown.

Sexual zonation generally is most pronounced in poor flowering individuals and years. In good flowering years and in response to successful induction treatments, the female zone extends inward and downward, sometimes to the base of the crown, and the male zone outward and upward (67,81,82, 118).

2.2. Mode and timing of differentiation

Owens and Blake (85) describe the mode and timing of differentiation of female and male strobili for many north temperate conifers (see Fig. 1). In Pinus, female strobili differentiate from lateral long-shoot primordia (potential branch buds) and male strobili differentiate from short-shoot primordia (potential needle fasicles). There are exceptions, however, such as rare cluster-cone variants in which female strobili differentiate from previously undetermined short-shoot primordia, or by the transition of an existing male strobili [see Rudolph et al. (121)]. In Tsuga, female strobili generally result from transition of an existing terminal vegetative apice, whereas male strobili may also occur this way but more frequently from differentiation of previously undetermined axillary primordia. Both types of differentiation also occur in Picea, but with the terminal mode predominating for females and the axillary mode predominating for males. In Pseudotsuga, strobili of both sexes differentiate from previously undetermined axillary primordia-- female strobili from distal primordia that would normally become vegetative branch buds, and male strobili from proximal primordia that would otherwise abort or remain latent.

The mode of strobilus differentiation in Abies is similar to that for Pseudotsuga, with the exception that vertical as well as horizontal orientation of an axillary primordia on the shoot can determine its developmental fate. Thus, female strobili tend to differentiate from primordia on the dorsal (upper) side of the diageotropic shoot, and male strobili from primordia in the ventral (lower) position. A similar situation exists for some species of Larix, although here strobili of both sexes generally occur on dwarf (short) shoots as a result of the transition of a previously vegetative terminal bud (61,77). Interestingly, gravimorphic treatments only appear to influence flowering and sex expression in these two genera. In L. leptolepis and L. kaemferi, tying horizontal branches in a vertical position promotes female flowering, whereas bending vertical branches downward promotes male flowering (61,62). Similarly, bending A. sibirica shoots so that the normally lower side faces upward causes axillary primordia in that position to differentiate as female rather than as male strobili (77).

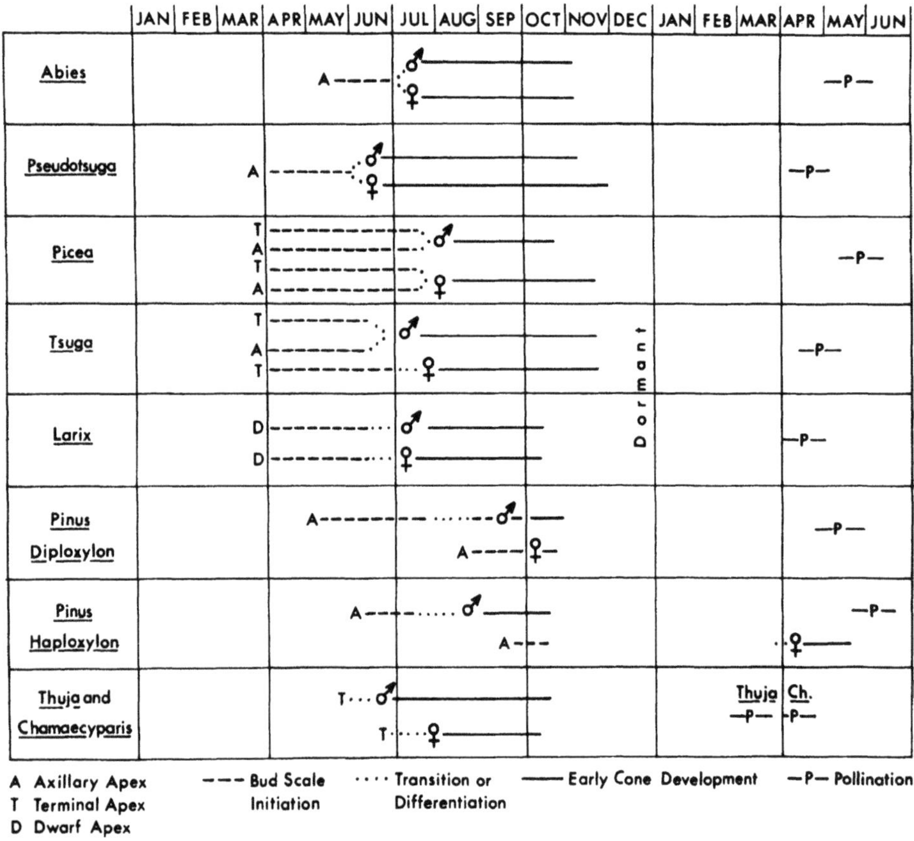

Figure 1. Times and modes of strobilus initiation in conifers [from Owens and Blake (85)].

Durand and Durand (21) concluded for angiosperms that "the male program is always expressed first, then the female one". Owens and Hardev (86) note that this applies to conifers as well if considered in a seasonal context (Fig. 1). This is very pronounced in Pinus, Tsuga and Cupressaceae and Taxodiaceae species, where on a given shoot male strobili are differentiated earlier in the season than female strobili, a fact confirmed by many timing trials using N fertilization and PGR treatments (16,48,81,82,127).

In Larix, Picea and Pseudotsuga, strobili of both sexes appear to differentiate at about the same stage of shoot development. Yet, even for these species the optimal time for promoting differentiation generally is earlier for male than female strobili (6,17,115,116,136). Owens and Hardev (86) speculate that this is because male strobili usually differentiate on weak shoots in the lower crown which complete their seasonal development earlier than those more vigorous, upper-crown shoots on which female strobili are typically borne. The importance of positional effect is indicated by studies on Tsuga heterophylla (118) and Larix species (6), in which the optimal time for the $GA_{4/7}$ promotion of female flowering was shown to be earlier for branches in the lower than upper crown.

2.3. Sex transformations

Although the reproductive structures of conifers are normally unisporangiate, bisporangiate and proliferated strobili occasionally occur in nearly all species (11,48,86,94,121,137). These range from mainly vegetative (Fig. 2A) to mainly female (Fig. 2B), and include all gradiations of male>female transformations (Fig. 2C). Female>male and male>vegetative transformations also occur but more rarely. Such transformations may occur spontaneously as a more or less normal event in young trees entering sexual maturity (11, 137), but more commonly in response to promotive stresses (3,25), pruning (48) and PGR treatments (48,69,94; S.D. Ross and R.F. Piesch, unpubl.).

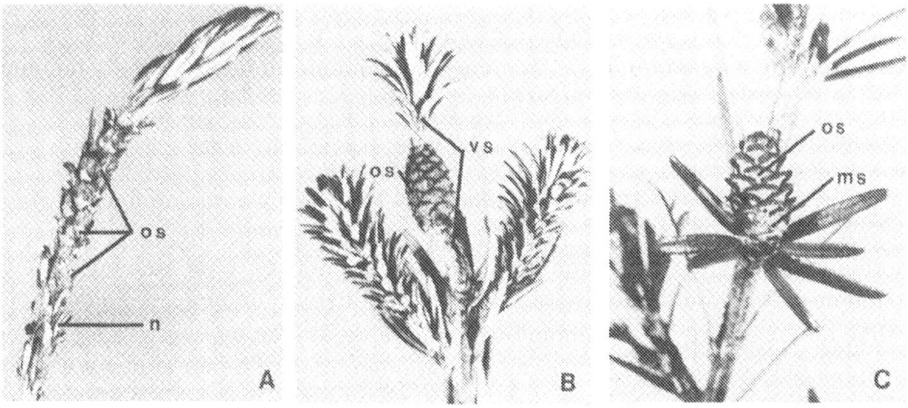

Figure 2. Examples of sex transformations produced by Tsuga heterophylla rooted cutting following treatment with $GA_{4/7}$. Fig. 2A. Rudimentary ovuliferous scales (os) interspersed among incompletely developed needles (n) along a vegetative shoot. Fig. 2B. Transformation of a vegetative shoot (vs) into a normal appearing female strobilus in which the apical meristem remained vegetative and produced a second growth flush. Fig. 2C. A bisporangiate cone in which megasporangiate sporophylla (ms) were initiated first followed by ovuliferous scales (os).

Sex transformations have long been a subject of anatomical interest, but their real significance is to illustrate the developmental plasticity of apical meristems during differentiation. Proliferated strobili are most likely to occur at the vegetative-female and vegetative-male transition zones in the tree crown, and bisporangiate cones in the male-female transition zone. Pharis and Morf (94) describe for Thuja plicata a bisporangiate strobili that started off female while on a vigorous shoot in the middle crown, but then became male (female development was arrested) as the branch became part of the lower crown.

Figure 2A shows a proliferated shoot that developed on a $GA_{4/7}$-treated Tsuga heterophylla. It began by initiating needle primordia (seen at the base), followed by megasporangiate primordia interspersed with needle primordia, and finally again only needle primordia. We attribute this unique development to a variable uptake of $GA_{4/7}$ throughout the prolonged differentiation period. Although $GA_{4/7}$ was applied by continuous stem injection, the injection hole rapidly became occluded and a new hole had to be drilled every 2 weeks over the 6-week treatment period [see Ross et al. (118)]. Thus the timing may have been such that the $GA_{4/7}$ dose was relatively low during the early and late stages of differentiation (resulting in vegetative development), but became high enough during the middle period to promote the formation (albeit sporadic) of ovuliferous scales (Fig. 2A). A similar situation may have existed for the shoot in Figure 2B, except that it was exposed to sufficient $GA_{4/7}$ for normal female strobilus development for a longer period midway through differentiation.

3. ROLE OF NUTRIENT STATUS AND VIGOR

3.1. Carbohydrates and nitrogen

Female strobili generally occur on vigorous shoots in the exposed upper crown and male strobili on suppressed, usually lower shoots that often but not always grow in shade. This has lead to the suggestion that a high photosynthetic rate favors female flowering, and conversely a low carbohydrate status favors male flowering (34,140). Although trees grown in dense shade may produce only male strobili (36,73,123), it is also well known that the production of both female and male strobili is maximized under conditions of wide tree spacing and associated high light intensities (34,70). The specific effect of light intensity on sex expression was investigated for Chamaecyparis obtusa under controlled-environment conditions (80). Nagao and Sasaki (80) confirmed that the threshhold light intensity required for initiation and development of male strobili (12 klux) was significantly lower than that for female strobili (21 klux), but also demonstrated that production of strobili of both sexes was greatest under the highest light intensity.

Silen (129) observed that enclosing Pseudotsuga menziesii shoots in white cloth bags (from 6 weeks before until 8 weeks after vegetative bud burst in spring) inhibited female flowering while increasing the production of male strobili. Temperatures, however, were up to 6.6°C higher in the bags, and Ross (110; unpubl.) found that comparably timed heat treatment in a polyethylene-covered house can have a similar effect on sex expression in this species. Changes in light quality may also be involved in the response to bagging. Hashizume (48) found that placing Pinus densiflora shoots in brown paper bags caused potential female strobili to apparently revert to male, whereas the reverse transformation occurred within red paper bags.

Removal of the photosynthetic apparatus by defoliation has stimulated male and inhibited female flowering in P. sylvestris (32), but a similar approach caused feminization of potential male strobili in P. thunbergii (48). A solely nutritional hypothesis also is difficult to reconcile with the observation that girdling, a treatment that may be expected to cause an accumulation of carbohydrates above the phloem blockage, will often preferentially promote male flowering (64,75,96,143). Thus, as with floral initiation/differentiation in general [see Ross and Pharis (115)], a direct morphogenic role for carbohydrates in sex expression of conifers remains unproven.

Wareing (140) has proposed that a low N status (i.e. high C/N ratio) favors male flowering, whereas a higher N status is necessary for production of female strobili. Reports of profuse male flowering in young, potted seedlings of Cryptomeria japonica (66), Cupressus arizonica (59) and Pinus radiata (132) in response to induced N deficiency are consistent with such a hypothesis. However, for all three species male flowering is more precocious than female, and in each of these studies the plant material was relatively young. Hence, a non-specific response of N deficiency on flowering per se-- as opposed to sex expression-- cannot be ruled out. Interestingly, in another study where C. japonica seedlings also received GA_3, NH_4-N promoted only male flowering and NO_3-N only female flowering (82).

There are reports that N fertilization has promoted only female flowering, with the sometimes inhibition of male flowering, in Pinus species (35,138) and Cryptomeria japonica (82). However, the treatment has promoted male flowering in other studies on Pinus (1,32,50,125), sometimes to a greater extent than female flowering (127). Further adding to the confusion is Mergan's (74) finding that application of liberal doses of NPK fertilizer increased the frequency of female>male transformations in P. griffithii x P. strobus seedlings. Schmidtling (127) noted that the optimal time for promotion of male flowering in P. taeda by ammonium-nitrate was earlier than that for promotion of female flowering, and thus timing of N treatment may partly account for these conflicting results.

Barnes and Bengtson (1) also questioned that a low shoot N content might favor male flowering in P. elliottii. They note that even though male flowering was inhibited by ammonium-nitrate treatment, there was no relationship among non-fertilized trees between the shoot content of N and production of male strobili. Similarly, in the dioecious angiosperm papaya (Carica papaya), N fertilization may increase the percentage of female plants and also causes sex reversal in male plants; yet mature leaves from male plants can have higher N contents than those from female plants (30). It thus appears that if the nutritional status of shoots has a role in sex expression it may be indirect through effects on bud vigor and/or hormonal status [see Ross and Pharis (115)].

3.2 Bud vigor

Among the first to notice the involvement of vigor in sex expression was Meehan [cited by Mergen (74)], who proposed that the sex of a strobilus is determined by the vigor of the branch on which it is formed. Tompsett (134) also concluded that there exists a causal relationship between the early vigor or meristematic activity of a primordial shoot and its developmental fate in Picea sitchensis. In that species and in many other Pinaceae conifers (86,108), the probable development of a primordial shoot appears to change as follows with decreasing early vigor: strong vegetative > female strobilus > intermediate vegetative > male strobilus > weak vegetative.

Exogenous GAs are known to stimulate subapical meristem activity in many higher plants (54), and Tompsett and Fletcher (136) suggested that this may be the mechanism by which $GA_{4/7}$ treatment promotes flowering in conifers. Thus, whether exogenously applied $GA_{4/7}$ promotes female or male flowering, or continued vegetative development in a primordial shoot, depends on the new vigor class into which the treated shoot is elevated.

Girdling, rootpruning and drought are all "stress" treatments that retard, at least temporarily, shoot elongation and bud development (41,46,62,89,90,142). As was previously noted for girdling, rootpruning and drought also tend to preferentially promote male flowering (10,23,117,120), except in young, vigorously growing trees where female flowering only may be increased (70,118,120). This is in agreement with Tompsett and Fletcher's (136) hypothesis, which predicts that stress treatments should preferentially promote female flowering in young tree where a majority of shoots fall into a "high vigor" category, but mainly male flowering in older trees where most of the shoots are of an intermediate vigor.

In Pseudotsuga menziesii, the promotion of flowering by calcium nitrate fertilization may be accompanied by either an enhanced (24) or a decreased (115) rate of shoot elongation, apparently depending on the tree's physiological status and dosage applied. Thus, given the variable influence of N fertilization on vegetative vigor, an inconsistent effect on strobilus sexuality is not surprising.

The variable response to top and branch pruning [see Ross and Pharis (115)] also seems amenable to explanation by a bud vigor hypothesis. Where only the main-stem terminal and leading shoots of vigorous upper-crown branches are pruned, production of female strobili on subtending branches is invariably depressed, even though these shoots are in the normal female region of the crown. However, the shoots are moderately vigorous, retaining the capacity to respond to the top pruning with greatly enhanced vegetative growth. And, when released from apical control, their very vigorous growth appears to predispose them to a reduced ability to differentiate female strobili. However, vigorous vegetative growth is not totally antithetical to flowering, since the $GA_{4/7}$ promotion of female flowering frequently is accompanied by enhanced vegetative bud and/or shoot elongation (46,60,108,142).

On the other hand, a severe top pruning which removes the upper third to half of live crown may actually increase the production of female strobili for the tree as a whole (71,83,111). In this instance the residual crown is comprised of many intermediate-vigor shoots in what was previously the female-male transition and male zones. Such shoots will respond to top pruning with only a moderate increase in vegetative vigor, and this apparently is sufficient to predispose them to differentiate female strobili.

Pollen production, in turn, may be depressed by this severe top pruning treatment, as the weaker shoots on which male strobili normally are borne are elevated to a vigor class where they remain vegetative or begin producing female strobili. But there exist numerous other shoots in the lowermost crown that are normally too suppressed to produce even male strobili. Top pruning can affect a portion of these, but an even more effective treatment appears to be a terminal pruning or disbudding of the lower branches themselves (48,73). A combination of annual top and side pruning has proved highly effective in increasing both female and male flowering potential in potted and field-grown Tsuga heterophylla orchard trees (111).

Despite its attractiveness in explaining the effects of various treatments on sex expression in conifers, there is little direct evidence for the bud vigor hypothesis. We (97,99,115,119) review elsewhere the evidence that GAs of a less polar nature play a direct morphogenic role in flowering of conifers, a role that is independent of any effect on bud vigor or nutritional status. Furthermore, it now appears that the flowering response to many cultural treatments may be, at least in part, hormonally mediated. Promotive drought, rootpruning, girdling, heat and nitrate fertilization treatments have all been shown to result in elevated endogenous levels of less polar GAs in shoots at time of strobilus differentiation (18,20,96,97,141).

4. ROLE OF PLANT GROWTH REGULATORS

4.1. Lessons from angiosperms

Sex expression in many angiosperms appears to be under strong control, and to involve the interactions between various classes of PGRs, including GAs, auxin, cytokinins, ethylene and abscissic acid (ABA) [see reviews by Vince-Prue (139), Pharis and King (92) and Sladky (130)]. While the analogy between angiosperms and gymnosperms with regard to sex expression is no doubt fraught with exceptions, there are many similarities and some valuable lessons for manipulating the sex of conifers to be learned from our experience with angiosperms. For example:

(i) A PGR may function either as an apparent male or female "sex hormone" in different species or even cultivars of the same species. GAs are quite variable in this regard (92,139), but differential responses have also been reported for cytokinins and ABA (27,28). High auxin and ethylene levels generally are associated with femaleness, although again there are exceptions (92,130).

(ii) The effect of a PGR on sex expression may vary with the plant's developmental stage (78). This is most common with monoecious species which initiate male and female flowers at different stages of ontogeny (92).

(iii) Timing of PGR application in relation to stage of floral development is important. For example, in corn the tassel may become pistillate, male sterile or remain male fertile as GA_3 treatment is applied progressively closer to the onset of microspore meiosis (45).

(iv) The female or male role of a PGR may be concentration dependent and also dependent on the presence of other applied PGRs and/or environmental factors, such as photoperiod, which may influence endogenous hormone levels (27,28,92,139).

(v) PGRs may influence the sex ratio directly through the sex of flowers initiated or indirectly, either through differential abortion of stamens or pistils, or by altering the growth and branching pattern in such a way as to favor the production of flowers of one sex over another (92).

These lessons learned from angiosperms no doubt apply to the study of PGRs in relation to sex expression in conifers. When applying PGRs or sampling tissues for endogenous contents the temporal and spatial patterns of sex expression for the species in question must be taken into account. Whereas usually the PGR is applied at only one concentration, a graded series of concentrations will probably be necessary to establish its effect on sex expression. It is furthermore necessary to examine the interactions between different PGRs (GAs and auxin especially), as sex expression may be controlled more by their balance than the absolute level of any one PGR. Photoperiod and temperature can also influence sex expression (discussed later), and should therefore be taken into account in designing treatments.

When assessing treatment responses it is just not sufficient to count the female and male strobili initiated. These should be closely examined for differential abortion (and the possible timing of its occurrence); for changes in their normal distribution along the shoot and within the crown; and for the occurrence, type and location of any sex transformations. Finally, close attention should be given to changes in growth rate and branching patterns that may indirectly influence the sex ratio.

Unfortunately, the above criteria are seldom met, and consequently the existing literature provides a very confused understanding of the potential role and practical uses of PGRs to control sex expression.

4.2 Gibberellins

As already noted, there is compelling evidence that GAs, especially those of a less oxidized or polar nature (e.g. GA_4, GA_5, GA_7, GA_9), play a direct morphogenic role in floral initiation/differentiation in conifers (17,20,92,93,96,97,99,115,119,141). A role for GAs in sex expression is also implied. In the Cupressaceae and Taxodiaceae, the threshhold concentration of GA_3 required to initiate male strobili is lower than that for female strobili, whose absolute number and proportion, however, increases with increasing concentration (47,62,63,81,82,93,94). In *Thuja plicata*, it appears as well that the minimum duration of GA_3 treatment required to induce male flowering is shorter than that for inducing females (S.D. Ross, unpubl.). Treatment with GA_3 was reported by Hashizume (48) to cause masculinization of female strobili in *Cryptomeria japonica* seedlings. However, timing of the GA_3 treatment can also determine whether male (June-Aug. application) or female (Aug.-Sept. application) flowering is primarily promoted in this species (47,82,122).

The role of GAs in sex expression of Pinaceae species is less clear. Bioassays of extracts from male versus female clones of *Picea abies* (53) and *Pinus sylvestris* (141), crown zones in *P. sylvestris* (56), and shoot regions in *Abies sibirica* (77) indicate that in these species, as in Cupressaceae and Taxodiaceae, male flowering is associated with higher levels of endogenous GAs as opposed to female flowering. However, the responses to exogenously applied $GA_{4/7}$ do not necessarily support this conclusion.

In several studies with *Pseudotsuga menziesii*, treatment with $GA_{4/7}$ alone has preferentially promoted female flowering (8,98,107,108,110), although there are exceptions (72,105). Timing of treatment has not affected the sex ratio in this species, although increasing the dosage of $GA_{4/7}$ can increase the proportion of male strobili initiated (108).

Treatment with $GA_{4/7}$ alone also has primarily promoted female flowering in *Tsuga heterophylla* (117,118). However, by using adjunct cultural treatments (e.g. drought, rootpruning or heat) plus $GA_{4/7}$ is it possible to obtain adequate male flowering (10,103,117,118). Male strobili differentiate 2-3 weeks before female strobili in this species (Fig. 1), and it is possible that a very early $GA_{4/7}$ treatment (as vegetative buds are beginning to swell) will preferentially promote male flowering (A. Colangeli, per. comm.).

The situation with regard to $GA_{4/7}$ effects on sex expression in *Pinus* is somewhat confused. In *P. sylvestris*, $GA_{4/7}$ treatment timed to the normal periods of male (early) and female strobilus differentiation (late) has primarily promoted that sex (16; O. Luukkanen and R. Hagqvist, per. comm.). The optimal $GA_{4/7}$ application times also differed for promotion of male and female flowering in accelerated-growth *P. banksiana* seedlings, although not in a manner that seemed to relate to the known timing of strobilus differentiation in this species (13). On the other hand, treatment with $GA_{4/7}$ alone has only promoted female flowering in P. taeda, regardless of age of trees, cultural conditions or timing and application dosage (42-44,112).

In Picea, exogenously applied $GA_{4/7}$ has tended to preferentially promote female flowering, both in seedlings and mature trees (19,64,68,70,100,135). However, it is possible that in many of the studies the $GA_{4/7}$ was applied past the optimal time for influencing male flowering (15,109,136), or was applied to that portion of the crown with an inherently high potential for producing female strobili. Using mature field-grown P. glauca trees, Marquard and Hanover (1984a), found that $GA_{4/7}$ promoted primarily female flowering when applied to the female-male transition zone but male flowering when applied to the lower, strictly male zone.

Bonnet-Masimbert (6) reported a similar differential response by upper- and lower-crown branches of Larix leptolepis and L. decidua to applied $GA_{4/7}$. However, both in his study and Katsuta's (55) on L. leptolepis, the $GA_{4/7}$ elicited a relatively greater female response for the tree as a whole, even though the treatment apparently was more optimally timed to influence male strobilus differentiation.

Given the experience with angiosperms (see above), it is not surprising that exogenously applied GAs can affect the sex ratio differently in different conifer species and even in different studies on the same species. Although it appears that GAs may preferentially promote male flowering in some conifers (e.g. Cupressaceae and Taxodiaceae) and female flowering in others (e.g. Pseudotsuga and Tsuga), there are yet others (e.g. Pinus, Picea and Larix) for which this PGR does not seem to specifically influence sex expression. Thus, it may be as Chalupka (14) suggests, that GAs function as a triggering agent to induce initial changes in the apical meristem which lead to reproductive as opposed to vegetative development. Whether the GA-promoted meristem differentiation occurs in a female or male direction is then a function of other factors, such as crown position, timing, photoperiod, and/or interaction of the GA with other PGRs.

4.3. Auxin and ethylene

Heslop-Harrison (51) notes that sex expression in many plants may be controlled by auxin gradients originating in the stem and branch apex during the course of normal differentiation. Ethylene, well as auxin, seem to have a relatively consistent role in sex expression in herbaceous and woody angiosperms (130,139)-- high contents of each being associated with femaleness. It is known that exogenously applied indoleacetic acid (IAA) can increase ethylene production in vegetative tissues of several higher plants, and the degree of stimulation increases with increasing auxin concentrations (52). This effect of exogenous auxin is apparently brought about by increasing the activity of ACC-synthase (ACC= 1-aminocyclopropane-1-carboxylic acid), an immediate biosynthetic precursor of ethylene. The auxin- increased ACC synthase activity can be synergized by a cytokinin (BA) and reduced by ABA (52). Hence, even the relatively clear-cut effect shown by auxin and ethylene may involve other PGRs.

Some work with Pinus species also has noted a high endogenous auxin content of shoots from the crown region that is normally female, relative to the region that is normally male (35,48), and similarly clones that tend to produce mainly female relative to male strobili (56,58,141). In Abies sibirica, Minina and Belyaev (76) found that the upper, female side of the diageotropic shoot had a higher endogenous auxin content than the lower, male side.

In the Cupressaceae and Taxodiaceae, exogenously applied auxins have generally but not always promoted femaleness (48,61,63). Ethrel, an ethylene-generating compound, also has had variable effects on the sex ratio when given with GA_3 (4,49). Bonnet-Masimbert (4) found with Cupressus arizonica that Ethrel stimulated female flowering under photoperiods (short day) favoring femaleness but males under long days favorable to male flowering. Nagao et al. (81) reported that Ethrel enhanced both the female and male response to GA_3 in Chamaecyparis obtusa.

Some early reports indicated that exogenously applied auxins will increase the proportion of female strobili produced in Pinus (48,93) and Larix (48). However, in subsequent studies on Pinaceae species, auxins alone generally have had little promotive effect on flowering (15,75,10,105).

On the other hand, auxins can modify the flowering response by some Pinaceae species to applied GAs. Contrary to the situation in Cupressaceae, and also the evidence for endogenous shoot contents (see above), there is some evidence that auxins [naphthaleneacetic acid (NAA) being more effective than IAA] can preferentially enhance the male response to $GA_{4/7}$ in Pseudotsuga menziesii (96,98,107) and Picea sitchensis (133). Treatment with NAA alone also has caused the reversion of female to male strobili in L. leptolepis and Pinus densiflora (48), and inhibited female flowering in P. banksiana (13). And, in P. taeda, branch applications of triiodobenzoic acid (TIBA), an inhibitor of polar auxin transport, slightly increased the production of female strobili (112).

These male-promotive results for auxin have not, however, been noted for Tsuga heterophylla (S.D. Ross and R.F. Piesch, unpubl.), or in all studies on Pseudotsuga (105), Pinus (13,65) or Picea (70,100). Frequently the auxin treatment resulted in depressed female and/or male flowering, which may be a phytotoxic response. Furthermore, in all these studies the auxin was tested at only one, or at most two concentrations, either or both of which may have been outside an optimal range for affecting sex expression.

4.4. Cytokinins

It has been noted that many of the treatments which promote flowering (e.g. rootpruning, drought and flooding of roots, high temperatures and excesses or deficiencies of N) could retard root growth. This has led to the suggestion (8,101) that actively growing roots export chemical substances inhibitory to flowering. Zaerr and Bonnet-Masimbert (146) present results for potted Pseudotsuga menziesii grafts which suggest that this inhibitory substance(s) may be a cytokinin. The promotion of flowering by $GA_{4/7}$ alone, and especially with flooding of roots, caused a marked decrease in endogenous (immunoassay active) cytokinin-like substances in shoots harvested during the period of strobilus differentiation. From a comparison of distal and proximal shoot segments, it also appeared that the threshhold concentration of cytokinin-like substance that was presumably inhibitory to male flowering was lower than that required to inhibit female flowering.

However, Bonnet-Masimbert and Zaerr in Chapter 2 present additional results where the relationship between endogenous cytokinin-like substances and flowering is less clear. Furthermore, in contrast to their trends for endogenous cytokinin-like substances, the synthetic cytokinin, benzyladenine (BA), when coadministered with GAs, has promoted female flowering in P. menziesii (113) but male flowering in Picea sitchensis (133). Thus, the picture for cytokinins and sex expression in conifers is by no means clear. The situation is not unlike that in some herbaceous angiosperms, where cytokinins and GAs both may be involved in sex expression (92,139).

5. ROLE OF PHOTOPERIOD AND TEMPERATURE

That photoperiod can influence sex expression in some Cupressaceae and Taxodiaceae conifers is well established (4,63,79-82,87,88). Long days (LD) generally maximize the absolute production of strobili of both sexes. But, whereas male flowering has a quantitative requirement for LD, shortening the day length usually increases the proportion of female strobili initiated. Thus, the control of sex expression by photoperiod generally is not absolute, although in one study with Cryptomeria japonica (79), treatment with GA_3 induced only female strobili under 8-h days and only male strobili under 24-h days.

In nature male and female strobili of Cryptomeria differentiate under LD and SD, respectively, whereas this apparently is not the case for all members of the Cupressaceae (87). In Chamaecyparis, Cupressus and Thuja, males are initiated under LD and increasing day length and female strobili also under LD, but when day length is decreasing. Thus, for these species sex expression may depend on the direction of change in day length and not the actual photoperiod (87). However, the importance of photoperiod (or its direction) in influencing sex expression in nature is still not clear. In field and greenhouse studies conducted under natural photoperiods on Thuja plicata (S.D. Ross unpubl.) and Chamaecyparis nootkatensis (R.C. Bower and S.D. Ross, unpubl.), GA_3 treatment became less effective in promoting both female and male flowering the progressively later the GA_3 was applied following flushing in spring. This is consistent with the observation (81) that LD maximize the production of strobili of both sexes in C. obtusa.

Giertych (31) has equated male strobilus initiation in Pinus with LD flowering behavior, and female flowering with SD flowering behavior. Evidence for this based on the southward and northward movement of provenances and clones, in which trees experience shorter and longer day lengths, respectively, in summer than in their natural range is inconclusive, however. In most such trials with Pinus species (29,124,126,128,144), the southward movement of provenances and clones has significantly increased not only female but also male flowering. This suggests a response to warmer, more favorable temperatures, rather than to changing photoperiod per se (34,128). In one comparison of 50 Pinus ponderosa var. scopulorum provenances from latitudes 32.3°N to 47.6°N (106), the northern provenances as a group produced more female strobili when planted in the middle of the range (lat. 41.0°N), whereas the southern provenances in the same plantation produced more male strobili, thus implying a photoperiod causality.

Kosinski and Giertych (57) found with Pinus sylvestris and Picea abies that increasing natural (white) light in the vacinity of the terminal bud apical dome using an optical fiber inserted through the bud scales increased the probability of that bud differentiating female strobili. The authors suggest that since bud scales absorb more red (R) than far-red (FR) (104), complete decay of the Prf to Pf form of phytochrome might not occur under the white light treatment rich in R, thereby signaling a long night (SD) situation. However, there may be an even more direct effect. FR enrichment of normal photosynthetically active radiation can stimulate GA metabolism (2,84), so that R (e.g. white light) enrichment may result in decreased GA hydroxylation. This could lead to the build up of less polar GAs, the precursors of polar GAs upon whose adequate availability flowering appears to depend (97,99).

Durzan et al. (1979) reported the inhibition of female flowering in field-grown Picea glauca trees using night-interrupted exposures to R via a laser during the period of strobilus differentiation. However, the fact that there was a year lag between treatment application and its effect on strobilus initiation raises questions about its photoperiodic basis. Furthermore, there was no mention of male flowering being affected by the LD treatment.

The issue is further complicated by the finding (37,91) that night interruptions of FR were more effective than R in mimicking the LD-enhancement of male strobili in GA_3-treated Cupressus arizonica seedlings. However, there is good precedent for FR being promotive of flowering in LD plants (26), and if FR were increasing overall GA content and metabolism within the plant (2,84), then it might be expected to interact with exogenous GA_3 application in an additive or synergistic way to increase flowering (males is this study on Cupressus). It may be noted, however, that in Chamaecyparis obtusa night breaks with R were more effective than with FR in promoting both female and male flowering (81), whereas in Cryptomeria japonica R increased only females and FR increased only males (82). For both species, blue-filtered light promoted only male flowering.

There is evidence from recent growth chamber experiments that photoperiod can influence sex expression in at least two Pinaceae conifers, Pinus contorta (60) and Tsuga heterophylla (103). In conjunction with $GA_{4/7}$ treatment, SD favored female flowering and LD male flowering in both species (photoperiods of 10 & 19.6 h for Pinus, 13 & 18 h for Tsuga). Each study also compared a warm and cool temperature regime (22/15°C & 15/8°C for Pinus, 30/30°C & 20/20°C for Tsuga). With Tsuga, female and male flowering were each significantly better (both photoperiods) under the warm regime. With Pinus, however, female flowering also was slightly better under the warm regime, whereas male flowering was significantly better in the cool regime. This contrasts with the situation in Cryptomeria, and to a lesser extent in Chamaecyparis, where the optimal temperature was higher for induction of male than female strobili (81,82).

Greenwood (40,42) has shown that when potted grafts of Pinus taeda which have overwintered in a greenhouse under continuous light are then exposed to naturally SD in spring, female flowering is dramatically enhanced. He attributes this not to a direct photoperiod effect on flowering, but rather to the premature setting of a terminal resting bud which occurs under the shortened photoperiod. Female strobilus differentiation in this species coincides with the end of shoot elongation, and the suggested effect of 'out-of-phase dormancy' is to enable sexual differentiation to occur before the onset of normal winter dormancy. However, the treatment enhances male flowering to an even greater extent, even though male strobilus differentiation normally occurs much earlier and, based on the results noted above for P. contorta (60), might be expected to be inhibited under the shortened photoperiod. Ross and Pharis (115) suggest that the effectiveness of out-of-phase dormancy may have a basis similar to drought, rootpruning and other stress treatments, which is to arrest vegetative growth and thereby allow the build up within the shoot of less polar GAs (possibly also nutrients) essential to flowering.

In Pseudotsuga menziesii, both female and male strobili are normally differentiated under a relative short but lengthening photoperiod in early-mid spring (Fig. 1). However, both sexes can also differentitate under relatively long but shortening photoperiods on lammas shoots which may form following a mid-summer drought (5). Bonnet-Masimbert (5) therefore concluded that flowering in this species depends on internal inductive conditions occurring at a specific stage of lateral primordium development, rather than at a precise time of the year.

Yokoyama and Asakawa (145) classed Larix leptolepis as day-intermediate on the basis that female strobili were produced under photoperiods of 12-16 h but not under shorter or longer photoperiods. Unfortunately, the authors used only a single clone which did not produce any males, and so their classification may be premature.

In the Cupressaceae, both the duration of the day period and intensity of light are important in a quantitative sense for male flowering especially (37,80,91,94). Thus, extending the photoperiod by low intensity supplemental light or by night breaks is not nearly as effective in promoting male flowering as is a longer exposure to high light. The implication is that a higher level of photosynthate (or GA) production favors quantitative increases in male over female flowering for Cupressaceae species. However, this is in contrast with the finding for Cryptomeria japonica (79) that initiation of female strobili requires a higher light intensity than initiation of male strobili.

Studies on Cupressus arizonica (37,91) also show that the endogenous GA content of shoots increases with day length, reaching maximum levels under continuous light of high intensity. Reconcilation of these promotive effects of LD/high endogenous GA on male flowering with the fact that increasing the concentration of exogenous GA_3 increases the female:male ratio (47,61,82,94) can only be accomplished if one involves a high dosage threshhold of GA for female flowering, or if one brings in other interacting factors such as auxin (e.g. high GA increases auxin production, and increased auxin:GA yields increased femaleness).

One further effect of photoperiod on sex expression bears mentioning. Following differentiation, there is generally the requirement for SD followed by LD for continued strobilus development through to anthesis (4,37,38,88,91,95,103). In the Cupressaceae and Taxodiaceae, it appears that neither female nor male strobili will develop without a SD treatment, which is most effective if administered at about 5 °C (4,94). However, the male and female strobili have differential requirements for photoperiod and/or cold, this requirement being satisfied sooner for males than for females (37,91,95). Also, in Pinus taeda, male strobili which are in a state of arrested development will resume growth under imposed LD, whereas female strobili require SD, unless GA_3 also is given under the LD (39).

6. CONCLUSIONS AND RECOMMENDATIONS FOR FURTHER RESEARCH

The way by which sex expression in conifers is regulated is poorly understood. This lack of understanding is as much a consequence of lack of critical study as a reflection of the complexity of the process. We suspect that, as is probably the case with with angiosperms, the process is under strong hormonal control, where balances between GAs, auxin, cytokinins and possibly inhibitors such as ABA are responsible for the observed gradients of sex expression along the stem, branch and shoot. However, the nature of this control and the possible involvement of non-hormonal aspects of shoot vegetative vigor and nutritional status remain obscure. For many conifers strobilus expression of sex depends on timing of the inductive stimulus, apparently in relation to stage of bud development. Nevertheless, there is evidence that for some species female and male flowering do have different photoperiod requirements.

Prospects for the operational control of sex expression in seed orchards to facilitate pollen and crop management also remain to be established. Yet, there are several potentially promising approaches, especially for use in indoor, containerized orchards where environmental and treatment conditions can be controlled. These include: (1) crown pruning to favor the production of specific types of shoots which have a high potential for differentiating either female or male strobili; (2) timing of treatments to coincide with the critical developmental stages for effecting female or male strobilus differentiation; (3) control of photoperiod for those species so responsive; and (4) manipulation of endogenous hormone balances through exogenous PGR applications under appropriate conditions noted in (1-3) above.

However, the successful development of these and other possible approaches for operational use requires that we have a better basic understanding of the natural control of sex expression. For example, it is still not clear when, during differentiation/early development of a strobilus, its sex is determined. There is some evidence from studies on Pseudotsuga menziesii (S.D. Ross, unpubl.) and Tsuga heterophylla (103) that sex expression can be influenced well after reproductive differentiation in lateral primordia is presumed to occur. Nor do we know if a particular effective treatment is influencing the sex ratio through a direct effect on floral initiation, through the subsequent differential development/abortion of previously initiated female and male strobili, or by altering branching and growth habits.

Answers to all of these questions will require the development of better techniques (anatomical, histochemical, and perhaps immunoassay) for detection of the very earliest stages of differentiation for female and male strobili. Finally, future research also should seek to characterize and quantify those positional, morphological and physiological attributes of shoots and buds which are constantly associated with sex expression. Such a characterization will aid, for example, in the practical development of sex-specific pruning regimes. It will also provide needed insight into designing sampling regimes for analysing PGRs and the carbohydrate and nitrogen status of shoots in relation to sex expression, and assessing flowering responses to different PGR and cultural treatments.

REFERENCES

1. Barnes RL and Bengtson GW (1968) Effects of fertilization, irrigation, and cover cropping on flowering and on nitrogen and soluble sugar composition of slash pine. For Sci 14:172-180
2. Beall, FD (1985) Phytochrome-regulated growth in beans (Phaseolus vulgaris cv. Kentucky Wonder): Investigations into the role of plant hormones. PhD. Thesis, Univ of Calgary, 168 p
3. Black TM (1961) Abnormalities of the reproductive system of Pinus contorta Loudon. Ann Bot 25:21-28
4. Bonnet-Masimbert M (1971) Induction florale precoce chez Cupressus arizonica et Chamaecyparis lawsoniana. Silvae Genet 20:82-90
5. Bonnet-Masimbert M (1979) Flowering on lammas shoots of Douglas-fir. Proc Flowering and Seed Development in Trees Symposium. Bonner F, ed. Washington, DC:US Gov Printing Office, pp 51-56
6. Bonnet-Masimbert M (1982) Effect of growth regulators, girdling, and mulching on flowering of young European and Japanese larches under field conditions. Can J For Res 12:270-279
7. Bonnet-Masimbert M (1987) Floral induction in conifers: a review on available techniques. For Ecol Manage. In press
8. Bonnet-Masimbert M, Delanzy P, Chanteloup G and Coupaye P (1982) Influence de l'etat d'activite des racines sur la floraison induite par des gibberellines 4 et 7 chez Pseudotsuga menziesii (Mirb.) Franco. Silvae Genet 31:178-183
9. Bower RC, Ross SD and Eastham AM (1986) Management of a western hemlock containerized seed orchard. Proc IUFRO Conf on Breeding Theory, Progeny Testing and Seed Orchards. Weir, RJ, ed. Raleigh NC:NC State Univ Press, pp 604-612
10. Brix H and Portlock FT (1982) Flowering response of western hemlock seedlings to gibberellin and water-stress treatments. Can J For Res 12:76-82
11. Caron GE (1984) Occurrence of sporangiate-vegetative structures and bisexual strobili in young plantations of black spruce. Proc 19th Mtg Can Tree Improvement Assoc, Part 1. Yeatman CW, ed. Ottawa:Can For Serv, pp 60
12. Cecich RA (1981) Applied gibberellin $A_{4/7}$ increases ovulate strobili production in accelerated growth jack pine seedlings. Can J For Res 11:580-585
13. Cecich RA (1983) Flowering in a jack pine seedlings seed orchard increased by spraying with gibberellin $A_{4/7}$. Can J For Res 13:1056-1062
14. Chalupka W (1980) Regulation of flowering in Scots pine (Pinus sylvestris L.) grafts by gibberellins. Silvae Genet 29:118-121
15. Chalupka W (1981) Influence of growth regulators and polythene covers on flowering of Scots pine and Norway spruce grafts. Silvae Genet 30:142-146
16. Chalupka W (1984) Time of $GA_{4/7}$ application may affect the sex of Scots pine flowers initiated. Silvae Genet 33:173
17. Chalupka W and Giertych M (1977) The effect of polyetheylene covers on the flowering of Norway spruce (Picea abies (L.) Karst.) grafts. Arboretum Kornickie 22:185-192
18. Chalupka W, Giertych M and Kopcewicz J (1982) Effect of polythene covers, a flower inducing treatment, on the contents of endogenous gibberellin-like substances in grafts of Norway spruce. Physiol Plant 54:79-82

19. Dunberg A (1980) Stimulation of flowering in Picea abies by gibberellins. Silvae Genet 29:51-53
20. Dunberg A, Malmberg G, Sassa T and Pharis RP (1983) Metabolism of tritiated gibberellins A_4 and A_9 in Norway spruce, Picea abies (L.) Karst. Plant Physiol 71:257-262
21. Durand R and Durand B (1984) Sexual differentiation in higher plants. Physiol Plant 60:267-274
22. Durzan DJ, Campbell RA and Wilson A (1979) Inhibition of female cone production in white spruce by red light treatment during night under field conditions. Env & Exp Bot 19:133-144
23. Ebell LF (1967) Cone production induced by drought in potted Douglas-fir. Can For Serv Bi-Monthly Res Notes 23:26-27
24. Ebell LF (1972) Cone-production and stem-growth responses of Douglas fir to rate and frequency of nitrogen fertilization. Can J For Res 2:327-338
25. Elliot DL (1979) Occurrence of bisexual strobiles on black spruce (Picea mariana [Mill.] B.S.P.) in the forest-tundra ecotone: Keewatin, Northwest Territories. Can J For Res 9:284-286
26. Evans LT (1971) Flower induction and the florigen concept. Ann Rev Plant Physiol 22:365-394
27. Friedlander M, Atsmon D and Galun E (1977) Sexual differentiation in cucumber: the effects of abscisic acid and other growth regulators on various sex genotypes. Plant & Cell Physiol 18:261-269
28. Friedlander M, Atsmon D and Galun E (1977) Sexual differentiation in cucumber: abscisic acid and gibberellic acid contents of various sex genotypes. Plant & Cell Physiol 18:681-691
29. Gansel CR (1973) Should slash pine seed orchards be moved south for early flowering? Proc 12th Southern For Tree Improvement Conf, pp 310-316
30. Ghosh, SP and Sen SP (1975) The modification of sex expression in papaya (Carica papaya L.). J Hort Sci 50:91-96
31. Giertych M (1967) Analogy of the differences between male and female strobiles in Pinus to the differences between long- and short-day plants. Can J Bot 45:1907-1910
32. Giertych M (1970) Generative development in scots pine (Pinus silvestris L.) grafts. Arboretum Kornickie 15:71-92
33. Giertych M (1976) Sexual expression in pines in relation to their ontogenetic development. Acta Hortic 56:99-104
34. Giertych M (1978) Role of light and temperature in flowering of forest trees. Proc 3rd IUFRO World Consultation on Forest Tree Breeding, Vol. 2. Camberra:CSIRO, pp 1013-1021
35. Giertych M and Forward DF (1966) Growth regulator changes in relation to growth and development of Pinus resinosa Ait. Can J Bot 44:717-738
36. Giertych M and Krolikowski Z (1978) Importance of bud insolation on female flower induction in pine (Pinus sylvestris L.). Arboretum Kornickie 23:161-169
37. Glenn JL (1973) Investigations concerning the physiology of gibberellin-induced growth and sexual differentiation phenomena in Arizona cypress (Cupressus arizonica Greene) seedlings. Ph.D. thesis, Univ. of Calgary, 134 p
38. Goo M (1968) Photoperiod and flowering of Pinus densiflora seedlings. Tokyo Univ For Misc Inf Rep No 17, pp 101-104
39. Greenwood MS (1977) The role of dormancy in the development of male and female strobili of loblolly pine. For Sci 23:373-375

40. Greenwood MS (1978) Flowering induced on young loblolly pine grafts by out-of-phase dormancy. Science 201:443-444
41. Greenwood MS (1980) Reproductive development in loblolly pine: I. The early development of male and female strobili in relation to the long shoot growth behavior. Am J Bot 67:1414-1422
42. Greenwood MS (1981) Reproductive development in loblolly pine. II. The effect of age, gibberellin plus water stress and out-of-phase dormancy on long shoot growth behavior. Am J Bot 68:1184-1190
43. Greenwood MS (1982) Rate, timing, and mode of gibberellin application for female strobilus production by grafted loblolly pine. Can J For Res 12:998-1002
44. Greenwood MS, O'Gwynn CH and Wallace PG (1979) Management of an indoor, potted loblolly pine breeding orchard. Proc 15th Southern For Tree Improvement Conf, pp 94-98
45. Hansen DJ, Bellman SK and Sacher RM (1976) Gibberellic acid-controlled sex expression of corn tassels. Crop Sci 16:371-374
46. Hare RC (1979) Promoting flowering in loblolly and slash pine with branch, bud, and fertilizer treatments. Proc Flowering and Seed Development in Trees Symposium. Bonner F, ed. Washington DC:US Gov Print Office, pp 112-121
47. Hashizume H (1960) The effect of gibberellin upon sex differentiation in Cryptomeria japonica strobiles. J Jap For Soc 42:176-180
48. Hashizume H (1973) Studies on flower bud formation, flower sex differentiation and their control in conifers. Bull Tottori Univ For No 7, 139 p
49. Hashizume H (1975) Interrelationship of gibberellin and ethrel in the flower bud formation and growth of conifers. Bull Tottori Univ For No 8, 10 p
50. Heidmann LJ, Rietveld WJ and Trujillo DP (1979) Fertilization increases cone production and diameter growth of a 55-year old ponderosa pine stand in Arizona. Proc Flowering and Seed Development in Trees Symposium. Bonner F, ed. Washington DC:US Gov Print Office, pp 197-205
51. Heslop-Harrison J (1963) Plant growth substances. In: Vistas in Botany, III. Turrill WB, ed. New York:Pergamon Press, pp 104-194
52. Imaseki H, Yoshi H and Tokada I (1982) Regulation of auxin-induced ethylene biosynthesis in plants. In: Plant Growth Substances, 1982. Wareing PF, ed. London:Academic Press, pp 259-268
53. Ivanis IY, Kyalina LV and Khokhlina EV (1979) Gibberellin-like substances in the needles of Norway spruce clones with different microstrobile abundance. Soviet Plant Physiol 26:266-271
54. Jones RL (1973) Gibberellins: their physiological role. Ann Rev Plant Physiol 24:571-598
55. Katsuta M (1981) Recent advances in the promotion of strobilus production by gibberellins in Japanese conifers. Proc XVII IUFRO World Congr, Div 2, Kyota Japan, pp 47-53
56. Kopcewicz J, Zatorska Z, Kulikowska H and Szczesniak T (1977) Endogenous growth regulators in embryonic shoots of Scots pine at the time of male and female flower primordia initiation. Acta Soc Bot Pol 44:119-128
57. Kosinski G and Giertych M (1982) Light conditions inside developing buds affect floral induction. Planta 155:93-94
58. Krulikowska H, Kopcewicz J, Zatorska Z and Szczesniak T (1978) Auxins and gibberellins in embryonic shoots of Scots pine in relation to flower sex differentiation. Acta Soc Bot Polon 4:403-409

59. Kuo CG (1973) Growth retardation and nutritional stress in relation to vegetative growth and reproductive differentiation of seedlings of Cupressus arizonica Greene. PhD thesis, Univ of Calgary, 157 p
60. Longman KA (1982) Effects of gibberellin, clone and environment on cone initiation, shoot growth and branching in Pinus contorta. Ann Bot 50:247-257
61. Longman KA (1985) Variability in flower initiation in forest trees. In: Attributes of Trees as Crop Plants. Cannell MGR and Jackson JE, eds. Inst Terrestrial Ecol, Monks Wood Experimental Station, Abbots Ripton, Huntingdon UK, pp 398-408
62. Longman KA (1986) Effects of growth substances on male and female cone initiation in conifers. Biol Plant 27:402-407
63. Longman KA (1987) Thuja. In: Handbook of Flowering, Vol 6. Halevy A, ed. Boca Raton FL:CRC Press. In press
64. Longman KA, Dick DM, Mugglestone M and Smith RI (1986) Effects of gibberellin A_{4+7} and bark-ringing on cone initiation in mature Picea sitchensis grafts. Tree Physiol 1:101-113
65. Luukkanen O and Johansson S (1980) Flower induction by exogenous plant hormones in Scots pine and Norway spruce grafts. Silva Fenn 14:95-105
66. Lyr H and Hoffmann G (1964) Uber den Einfluss der mineralsalzernahrung auf die Fruhfruktifikation von Cryptomeria japonica (L.E.D.) Don. Flora Jena 155:189-208
67. Marquard RD and Hanover JW (1984) Sexual zonation in the crown of Picea glauca and the flowering response to exogenous $GA_{4/7}$. Can J For Res 14:27-30
68. Marquard RD and Hanover JW (1984) Relationship between gibberellin $GA_{4/7}$ concentration, time of treatment, and crown position on flowering of Picea glauca. Can J For Res 14:547-553
69. Marquard RD and Hanover JW (1984) The effect of shade on flowering of Picea glauca. Can J For Res 14:830-832
70. Marquard RD and Hanover JW (1985) Floral response of Picea glauca to gibberellin $A_{4/7}$, naphthaleneacetic acid, root-pruning, and biennial treatment. Can J For Res 15:743-746
71. Matheson AC and Willcocks KW (1976) Seed yield in a radiata pine seed orchard following pollarding. NZ J For Sci 6:14-18
72. McMullan EE (1980) Effect of applied growth regulators on cone production in Douglas-fir, and relation of endogenous growth regulators to cone production capacity. Can J For Res 10:405-422
73. Melchior von GH (1962) Weitere Untersuchungen zur Forderung der Blutenbildung an Kiefern durch Ruchschnitt. Silvae Genet 11:11-15
74. Mergen F (1963) Sex transformation in pine hybrids. For Sci 9:258-262
75. Mikami S, Asakawa S, Itsuka M, Yokoyama T, Nagao A, Takehana S and Kaneko T (1979) Flower induction in Japanese larch, Larix leptolepis Gord. Bull For & For Prod Res Inst (Ibaraki Jap) No 307, pp 9-24
76. Minina EG and Belyaev VV (1972) Study of growth substances in connection with dorsiventrality and sexualization of shoot in Siberian fir (Abies sibirica Ledeb.). Soviet Plant Physiol 19:658-662
77. Minina EG and Tret'yakova IN (1979) Significance of phytohormones for formation of female strobiles by Siberian fir (Abies sibirica Ledeb.) in a gravitational field. Soviet Plant Physiol 26:855-858
78. Mohan RHY and Jaiswal VS (1972) Induction of male flowers on female plants of Cannabis sativa by gibberellins and its inhibition by abscisic acid. Planta 105:263-266

79. Nagao A and Sasaki S (1981) Effects of temperature and light on promotion of strobilus of Cryptomeria japonica and Chamaecyparis obtusa. Proc XVII IUFRO World Congr, Symposium on Flowering Physiology. Krugman SL and Katsuta M, eds. Tokyo:Jap For Tree Breeding Assoc, pp 103-114
80. Nagao A and Sasaki S (1985) Effects of light treatments on floral initiation of Chamaecyparis obtusa seedlings. Bull For & For Prod Res Inst (Ibaraki, Jap) No 332, pp 39-60
81. Nagao A, Sasaki S and Pharis RP (1987) Flowering of Chamaecyparis. In: Handbook on Flowering, Vol 6. Halevy A, ed. Boca Raton FL:CRC Press. In press
82. Nagao, A, Sasaki S and Pharis RP (1987) Flowering of Cryptomeria. In: Handbook on Flowering, Vol 6. Halevy A, ed. Boca Raton FL:CRC Press. In press
83. Nienstaedt H (1981) Top pruning white spruce seed orchard grafts. Tree Planter's Notes 32:9-13
84. O'Brien T, Beall FD and Smith H (1985) De-etiolation and plant hormones. In: Encyclopedia of Plant Physiology, Vol 11 (NS). Pharis RP and Reid DM, eds. Berlin:Springer-Verlag, pp 282-311
85. Owens JN and Blake MD (1985) Forest tree seed production. A review of literature and recommendations for future research. Can For Serv Inf Rep Pl-X-53, 161 p
86. Owens JN and Hardev V (1987) Gymnosperms. In: Sex Differentiation in the Plant Kingdom. Durant B and Durand R, eds. Boca Raton FL:CRC Press. In press
87. Owens JN and Molder M (1977) Cone induction in yellow cypress (Chamaecyparis nootkatensis) by gibberellin A_3 and the subsequent development of seeds within the induced cones. Can J For Res 7:605-613
88. Owens JN and Pharis RP (1971) Initiation and development of western red cedar cones in response to gibberellin induction and under natural conditions. Can J Bot 49:1165-1175
89. Owens JN, Webber JE, Ross SD and Pharis RP (1985) Interaction between gibberellin $A_{4/7}$ and root-pruning on the reproductive and vegetative process in Douglas-fir. III. Effects on anatomy of shoot elongation and terminal bud development. Can J For Res 15:354-364
90. Owens JN, Webber JE, Ross SD and Pharis RP (1986) Interaction between gibberellin $A_{4/7}$ and root-pruning on the reproductive and vegetative processes in Douglas-fir. IV. Effects on lateral bud development. Can J For Res 16:211-221
91. Pharis RP (1974) Precocious flowering in conifers: the role of plant hormones. In: Toward the Future Forest: Applying Physiology and Genetics to the Domestication of Trees. Ledig FT, ed. Yale Univ School of For & Envir Studies Bull No 85, pp 51-80
92. Pharis RP and King RW (1985) Gibberellins and reproductive development in seed plants. Ann Rev Plant Physiol 36:517-568
93. Pharis RP and Kuo CG (1977) Physiology of gibberellins in conifers. Can J For Res 7:299-325
94. Pharis RP and Morf W (1968) Physiology of gibberellin-induced flowering in conifers. In: Biochemistry and Physiology of Plant Growth Substances. Wightman F and Setterfield S, eds. Ottawa:Runge Press, pp 1341-1356
95. Pharis RP and Morf W (1970) Sexuality in conifers: effects of photoperiod and gibberellin concentration on the sex of gibberellin-induced strobili of western red cedar. Zesz Nauk Uniw Mikolaja Kopernika Toruniu Nauki Mat Przyr Biol 13:85-89

96. Pharis RP and Ross SD (1986) Flowering of Pinaceae family conifers with gibberellin $A_{4/7}$ mixture: how to accomplish it, mechanisms and integration with early progeny testing. In: Conifer Tree Seed in the Inland Mountain West Symposium. Shearer RC, ed. USDA For Serv Gen Tech Rep INT-203, pp 171-179
97. Pharis RP and Ross SD (1986) Hormonal promotion of flowering in Pinaceae family conifers. In: Handbook of Flowering, Vol. 5. Halevy A, ed. Boca Raton FL:CRC Press, pp 269-286
98. Pharis RP, Ross SD and McMullan EE (1980) Promotion of flowering in the Pinaceae by gibberellins. II. Seedlings of Douglas fir. Physiol. Plant 50:119-126
99. Pharis RP, Webber JE and Ross SD (1986) The promotion of flowering in forest trees by gibberellin $A_{4/7}$ and cultural treatments: a review of the possible mechanisms. For Ecol Manage. In press
100. Pharis RP, Tomchuk D, Beall FD, Rauter RM and Kiss G (1986) Promotion of flowering in white spruce (Picea glauca) by gibberellin $A_{4/7}$, auxin (naphthaleneacetic acid), and the adjunct cultural treatments of girdling and Ca(NO3)2 fertilization. Can J For Res 16:340-345
101. Philipson JJ (1983) The role of gibberellin $A_{4/7}$, heat and drought in the induction of flowering in Sitka spruce. J Exp Bot 34:291-302
102. Pollard DFW and Portlock FT (1981) Effects of temperature on strobilus production in gibberellin-treated seedlings of western hemlock. Can For Serv Res Notes 1:21-22
103. Pollard DFW and Portlock FT (1984) The effects of photoperiod and temperature on gibberellin $A_{4/7}$ induced strobilus production of western hemlock. Can J For Res 14:291-294
104. Pukacki P and Giertych M (1982) Seasonal changes in light transmission by bud scales of spruce and pine. Planta 154:381-383
105. Puritch GS, McMullan EE, Meagher MD and Simmons CS (1979) Hormonal enhancement of cone production in Douglas-fir seedlings and grafts. Can J For Res 9:193-200
106. Read RA and Sprackling JA (1983) Flowering in a ponderosa pine provenance plantation in eastern Nebraska. USDA For Serv Res Note RM-423, 6 p
107. Ross SD (1976) Differential flowering responses by young Douglas-fir grafts and equi-sized seedlings to gibberellins and auxin. Acta Hortic 56:163-168
108. Ross SD (1983) Enhancement of shoot elongation in Douglas-fir by gibberellin $A_{4/7}$ and its relation to the hormonal promotion of flowering. Can J For Res 13:986-994
109. Ross SD (1985) Promotion of flowering in potted Picea engelmannii (Perry) grafts: effect of heat, drought, gibberellin $A_{4/7}$ and their timing. Can J For Res 15:618-624
110. Ross SD (1986) Temperature influences on reproductive development in conifers. Proc Forest Climate '86: Symposium on Climate Applications in Forest Renewal and Forest Production, 17-20 Nov 1986, Orillia, Ont. In Press
111. Ross SD and Eastham AM (1986) Seed orchard management research. Proc 20th Mtg Can Tree Improvement Assoc, Part 1. Yeatman CW and Boyle TJB, eds. Ottawa:Can For Serv, pp 199-202
112. Ross SD and Greenwood MS (1979) Promotion of flowering in the Pinaceae by gibberellins II Grafts of sexually mature and immature Pinus taeda L. Physiol Plant 45:207-210

113. Ross SD and Pharis RP (1976) Promotion of flowering in the Pinaceae by gibberellins. I Sexually mature, non-flowering grafts of Douglas-fir. Physiol Plant 36:182-186
114. Ross SD and Pharis RP (1982) Recent developments in enhancement of seed production in conifers. Proc 18th Mtg Can Tree Improvement Assoc, Part 2. Pollard DFW, Edwards DGW and Yeatman CW, eds. Ottawa:Can For Serv, pp 26-38
115. Ross SD and Pharis RP (1985) Promotion of flowering in tree crops: different mechanisms and techniques, with special reference to conifers. In: Attributes of Trees as Crop Plants. Cannell MGR and Jackson JE, eds. Inst Terrestrial Ecol, Monks Wood Experimental Station, Abbots Ripton, Huntingdon UK, pp 383-397
116. Ross SD and Pharis RP (1986) Status of flowering in conifers: a constraint to tree improvement? Proc 20th Mtg Can Tree Improvement Assoc, Part 2. Caron F, Corriveau AG and Boyle TJB, eds. Ottawa:Can For Serv, pp 11-28
117. Ross SD, Eastham AM and Bower RC (1986) Potential for container seed orchards. Proc Conifer Tree Seeds in the Inland Mountain West Symposium. Shearer RC, ed. USDA For Serv Gen Tech Rep NT-203, pp 180-186
118. Ross SD, Piesch RF and Portlock FT (1981) Promotion of cone and seed production in rooted ramets and seedlings of western hemlock by gibberellins and adjunct cultural treatments. Can J For Res 11:90-98
119. Ross SD, Pharis RP and Binder WD (1983) Growth regulators and conifers: their physiology and potential uses in forestry. In: Plant Growth Regulating Chemicals. Nickell LG, ed. Boca Raton FL:CRC Press, pp 35-78
120. Ross SD, Webber JE, Pharis RP and Owens JN (1985) Interaction between gibberellins $A_{4/7}$ and rootpruning on the reproductive and vegetative process in Douglas-fir. I. Effects on flowering. Can J For Res 15:341-347
121. Rudolph TD, Wheeler, NC and Dir NK (1986) Cone clusters in jack pine. Can J For Res 16:1180-1184
122. Saito T and Kawasaki H (1984) Studies on flowering characteristics and their heredity in Cryptomeria japonica D Don. Bull For & For Prod Res Inst Iberaki, Japan) No 328, pp 17-41
123. Sarvas R (1962) Investigations on the flowering and seed crop of Pinus sylvestris. Inst For Fenn 53:1-198
124. Sarvas R (1970) Establishment and registration of seed orchards. Folia For Fenn 89
125. Schmidtling RC (1971) Cultivating and fertilizing stimulate precocious flowering in loblolly pine. Silvae Genet 20:220-221
126. Schmidtling RC (1979) Southern loblolly pine seed orchards produce more cone and seed than do northern orchards. Proc Flowering and Seed Development in Trees Symposium. Bonner F, ed. Washington DC:US Gov Print Office, pp 177-186
127. Schmidtling RC (1983) Fertilizer timing to increase flowering in eastern white pine (Pinus strobus L). Proc 3rd North Central Tree Improvement Conf, pp 184-189
128. Schmidtling RC (1984) Planting south of origin increases flowering in shortleaf (Pinus echinata Mill) and virginia pines (P. virginiana Mill). Silvae Genet 33:140-144
129. Silen RR (1973) First- and second-season effect on Douglas-fir cone initiation from a single shade period. Can J For Res 3:528-534

130. Sladky Z (1986) The role of growth regulators in the differentiation of flowers and inflorescences. Biol Plant 28:31-37
131. Sweet GB and Krugman SL (1978) Flowering and seed production problems and a new concept of seed orchards. Proc 3rd IUFRO World Consultation on Forest Tree Breeding, Vol 2. Camberra:CSIRO, pp 749-759
132. Sweet GB and Will GM (1965) Precocious male cone production associated with low nutrient status in clones of Pinus radiata. Nature 206:739
133. Tompsett PB (1977) Studies of growth and flowering in Picea sitchensis (Bong) Carr. 1. Effects of growth regulator applications to mature scions on seedling rootstocks. Ann Bot 1171-1178
134. Tompsett PB (1978) Studies of growth and flowering in Picea sitchensis (Bong) Carr. 2. Initiation and development of male, female and vegetative buds. Ann Bot 42:889-900
135. Tompsett PB and Fletcher AM (1977) Increased flowering of Sitka spruce (Picea sitchensis (Bong) Carr) in a polythene house. Silvae Genet 26:84-86
136. Tompsett PB and Fletcher AM (1979) Promotion of flowering on mature Picea sitchensis by gibberellin and environmental treatments. The influence of timing and hormonal concentration. Physiol Plant 45:112-116
137. Tosh KJ and Powell GR (1986) Proliferated, bisporangiate, and other atypical cones occurring on young, plantation-grown Larix laricina. Can J Bot 64:469-475
138. van Buijtenen JP (1966) The effect of spacing, fertilization, and cultivation on flowering and seed production in loblolly pine. Proc 8th Southern Conf For Tree Breeding, pp 141-146
139. Vince-Prue D (1975) Photoperiodism in Plants. London:McGraw Hill, 444 p
140. Wareing PF (1958) Reproductive development in Pinus sylvestris. In: The Physiology of Forest Trees. Thimann KV, ed. New York:Ronald Press, pp 643-654
141. Wesoly W (1985) Effect of girdling on flowering and on levels of endogenous growth regulators in embryonic shoots of Scots pine grafts (Pinus silvestris L). Acta Physiol Plant 7:171-179
142. Webber JE, Ross SD, Pharis RP and Owens JN (1985) Interaction between gibberellin $A_{4/7}$ and root-pruning on the reproductive and vegetative process in Douglas-fir. II. Effects on growth. Can J For Res 15:348-353
143. Wheeler NC, Cade SC, Masters CJ, Ross SD, Keeley JW and Hsin LY (1985) Girdling: a safe, effective and practical treatment for enhancing seed yields in Douglas-fir seed orchards. Can J For Res 15:505-510
144. Ying CC, and Illingworth K (1986) Variation in cone and seed production from natural stands, plantations and clonal-banks of lodgepole pine. Proc Conifer Tree Seed in the Inland Mountain West Symposium. Shearer RC, ed. USDA For Serv Gen Tech Rep INT-203, pp 191-199
145. Yokoyama T and Asakawa S (1973) Effect of photoperiod on female strobili formation in Larix leptolepis (Sieb et Zucc) Gordon. J Jap For Sci 55:388-393
146. Zaerr JB and Bonnet-Masimbert M (1987) Cytokinin level and flowering in Douglas-fir. For Ecol Manage. In press
147. Zimmerman RH, Hackett WP and Pharis RP (1985) Hormonal aspects of phase change and precocious flowering. In: Hormonal Regulation of Development. III. Role of Environmental Factors. Encyclopedia of Plant Physiology (NS), Pharis RP and Reid DM, eds. Heidelberg:Springer-Verlag, pp 79-115

Plant Growth Regulation 6: 61–93

4. THE ROLE OF PLANT GROWTH REGULATORS IN THE GERMINATION OF FOREST TREE SEEDS

C.L. LEADEM
Research Scientist, Research Branch, British Columbia Ministry of Forests and Lands, 1320 Glyn Road, Victoria, B.C. V8Z 3A6

ABSTRACT
To establish a role for PGRs in the germination of forest tree seeds, it must be demonstrated that PGRs are involved in the mechanisms that control the induction of and release from dormancy, and also that these mechanisms are operative in the seeds of woody plants. Four currently known concepts of seed dormancy are reviewed, citing the published evidence relating PGRs to tree seed germination. The germination of tree seeds appears to be controlled by a variety of external and internal factors. PGRs figure prominently among these factors, but the range of mechanisms by which such control is mediated may vary considerably--from physical to metabolic. Although significant support can be found for PGR involvement in the regulation of angiosperm seed dormancy, the evidence for gymnosperms is still inconclusive. This is primarily because our understanding of the control of germination--by PGRs or other means--is generally unsatisfactory. However, recent advancements in PGR technology, together with a more holistic concept of PGR action in biological systems, signal a more favourable climate and a revitalized approach to the investigation of the role of PGRs in plant growth and development.

1. INTRODUCTION

An impressive literature exists on the topic of hormonal involvement in seed dormancy and germination. In the past 15 years a number of excellent reviews have been written attempting to consolidate this body of knowledge and uncover the mechanisms controlling germination of seeds (23,51,64,121,139,140). However, most plant growth regulator (PGR) research has focussed on herbaceous species, and thus only limited attention has been paid to the germination of forest tree seeds. Of the relatively small number of papers that have been published on tree seeds, the majority are concerned with angiosperms, usually those that bear edible fruits or nuts. The strength of the angiosperm literature lies in the fact that extensive studies have been performed on seeds of the same genera, notably Corylus, Pyrus, Acer, and Fraxinus. These collaborative efforts have advanced our understanding of hormones in broad-leaved trees, but notable deficiencies still exist in the gymnosperm literature. To compensate for this imbalance, emphasis will be given in this review to conifer seeds, although frequent references will also be made to the extensive body of knowledge relating to angiosperms.

2. CHARACTERIZATION OF DORMANCY

Many seeds do not germinate when placed under conditions which are normally regarded as favourable to germination, namely, an adequate water supply, a suitable temperature, and the normal gaseous composition of the atmosphere. Such seeds are viable but said to be in a state of dormancy because they can only be induced to germinate by special treatments. Forest tree seeds almost always exhibit some degree of dormancy. In the prescriptions for international certified testing of tree seeds, 70% of the 178 species listed required dormancy-breaking treatment (44). With such widespread occurance of dormancy, the role of PGRs in the germination of forest tree seeds cannot be fully appreciated without an examination of hormonal involvement in the dormancy process.

Attempts have been made to classify states of dormancy; one widely accepted scheme distinguishes exogenous and endogenous causes of dormancy (79). Among the exogenous types are physical, chemical and mechanical dormancy. Physical constraints to germination are due to the impermeability of seed coats to water or oxygen, whereas chemical dormancy is due to the presence of germination inhibitors. Mechanical dormancy is caused by the physical restraint of the seed coat, pericarp, megagametophyte, or by a combination of these parts. Dormancy due to endogenous factors can be morphological or physiological. Morphological dormancy is due to seed immaturity, and is most frequently treated by employing "warm" stratification (15 to 20°C) to enhance embryo development. Seeds also exhibit physiological dormancy in which metabolic requirements, usually initiated by light and cold temperatures, must be satisfied for the embryo to germinate. Physiological dormancy, the type most frequently encountered in forest tree seeds, can vary in degree from deep dormancy, requiring very long periods of chilling at 2 to 5°C (cold stratification), to non-deep where only minimal treatment is needed for germination. In many cases, combinations of the above dormancy types occur and several treatments are needed to release seeds from dormancy. For example, "warm" stratification followed by cold stratification may be required for seeds which exhibit both morphological and physiological dormancy.

3. MAJOR CLASSES OF PLANT GROWTH REGULATORS INVOLVED IN SEED GERMINATION

3.1. Gibberellins (GAs)

Gibberellins actively stimulate seed germination of many species of angiosperms (Table 1) and gymnosperms (Table 2). Over 60 different GAs have been identified (43), but those most frequently used exogenously in forest tree seeds are GA_3, GA_4, and GA_7. Dormancies with chilling and light requirements are often overcome by GAs, and synergistic interactions between phytochrome and applied GA have been cited (see Section 4.3). Increases in endogenous GA are usually, although not always, coincident with the termination of dormancy in seeds undergoing dormancy-releasing treatments such as stratification (119) and light (120).

Although germination of seeds is frequently associated with increased GA levels, it is not clear at what stage of the germinatrion process this increase occurs. It can only be stated that exogenous GA relieves dormancy and is an endogenous component of many tree seeds--a priori evidence for a significant role for GA at some point in the termination of dormancy, or the initiation of germination.

3.2. Cytokinins (CK)

Cytokinins promote growth and differentiation in cultured callus tissue, a feature utilized in their bioassay. Like other PGRs, they are implicated in many physiological phenomena as well as seed dormancy and germination. Applied CK usually displays low activity in dormancy and germination control

Table 1. Effects of plant growth regulators on germination of forest tree seeds (Angiosperms).

A. GIBBERELLINS (GA)

Species	Source of PGR	Effect	Comments	Citation
Acer platanoides	Endog	GA ↑	during strat and germ	92
A. platanoides	Exog	chilling period ↓ by GA	greater ↑ by CK	93
A. pseudoplatanus	Exog	GA ↑ cotyledon expansion no germ with GA	expansion inhib by exog ABA unstrat seeds	95
A. pseudoplatanus	Endog	no change in GA levels	5°C and 20°C	145
A. pseudoplatanus	Exog	no effect on germ	CK ↑ germ, endog ABA	147
A. saccharum	Exog	GA, CK effect in phase 2	no effect phase 1, ↓ during phase 3	109
A. saccharum	Endog	[GA] ↑	40+d strat, [ABA] ↓, [CK] ↑	142

ABBREVIATIONS FOR TABLES 1,2:

ABA	abscisic acid	IAA	indole acetic acid	solns	solutions
CK	cytokinin	incorp	incorporate	spp	species
d	day	inhib	inhibit, inhibitor	stim	stimulation, stimulate
endog	endogenous	morphol	morphological	strat	stratification
exog	exogenous	mo	month(s)	temp	temperature
GA	gibberellin	physiol	physiological	wk	week
germ	germination	rev	reverses	yr	year
h	hour(s)		increase		decrease

Species	Source of PGR	Effect	Other treatments	Citation
A. saccharum	Exog	↑ germ	remove pericarp, prick testa	143
A. saccharum	Endog	[GA]↑ , then [GA] ↓	40d strat, then further chilling	144
A. tartaricum	Exog	↑ germ rate GA no effect	excised embryos 2 mo warm strat	78
Fagus sylvatica	Exog	↑ germ	remove pericarp	12
Fraxinus americana	Exog	↑ germ	unstrat seeds	65
F. americana	Endog Exog	high GA GA ↑ germ	1-yr old seeds (low in old seeds) overcame high ABA in old seeds	116
F. americana F. ornus	Exog	GA + CK ↑ germ	rev ABA effect on excised nondormant embryos	113
F. excelsior	Endog	antagonized by ABA		135
F. excelsior	Exog	remove morphol immaturity	warm strat, but seeds also need cold strat to break physiol dormancy	141
Liquidambar spp	Exog	↑ germ	substituted for 30-90 d strat	17
Nothofagus obliqua N. procera	Exog	↑ germ ↑ embryo axis	100% germ for treated seeds 20% for controls	104
Tilia platyphyllos	Endog	↑ GA levels	Strat, GAs changed from bound to free	77
Carya, Fraxinus, Liquidambar, Liriodendron Platanus, Quercus	Exog	↑ germ rate	soaking at 23°C better than 3°C 48 and 96 h soaks better than 12 and 24 h	11

Species	Source of PGR	Effect	Other treatments	Citation
Acer, Euonymus	Exog	↑ germ	seeds with shallow dormancy	80
Fraxinus		GA no effect	seeds requiring cold strat	
		slight effect	during warm strat	
Morus, Fraxinus,	Exog	↑ germ	seeds grown on paper moistened	112
Betula, Caragana, Lonicera		↓ germ	with GA solns	
B. CYTOKININ (CK)				
Acer platanoides	Endog	CK ↑ during early strat	↓ late strat, ↑ during germ	92
A. platanoides	Exog	embryo, no response to CK	phase 1, responds only to chilling	93
		CK ↓ chilling period	phase 2	
A. platanoides	Endog	free and bound CK ↑	first 20 d at 5°C	48, 49
A. pseudoplatanus		CK ↓	20-60 d at 5°C	
A. pseudoplatanus	Endog	CK ↑, then CK ↓	during strat	30
		germ ↑ when CK ↓	at warm temp	
A. pseudoplatanus	Exog	CK ↑ radicle elongation,	elongation inhib by exog ABA	95
		stim germ unstrat seeds	exog ABA did not inhibit	
A. pseudoplatanus	Exog	↑ germ	can overcome endog ABA inhibition	147
A. pseudoplatanus	Endog	no change in CK	at 5°C or 20°C	145
A. saccharum	Exog	CK + GA affect embryo axis	phase 2	109
		no effect	phase 1	
		levels decrease	phase 3	
A. saccharum	Endog	[CK] ↑, [CK] ↓	20 d chilling, >20d chilling	133

Species	Source of PGR	Effect	Other treatments	Citation
A. saccharum	Endog	[CK] ↑	20-40d chilling, [ABA] ↓ 0-20, [GA] ↑ 40+d	142
A. saccharum	Exog	↑ germ	remove pericarp, prick testa	143
A. saccharum	Endog	[CK] ↑ , [CK] ↓	strat for 20 d, further chilling	144
Fraxinus americana	Exog	99% germ with zeatin	excised embryo	38
F. americana F. ornus	Exog	CK + GA ↑ germ	rev effect of ABA-inhibited germ of excised nondormant embryos	113
F. americana	Exog	zeatin ↑ germ	dormant embryos, ABA antagonistic to zeatin	131
C. INHIBITORS (ABA)				
Acer platanoides	Endog	ABA ↓ during strat	still significant amount ungerm seeds	92
A. pseudoplatanus	Exog	↓ radicle, cotyledon growth		95
A. pseudoplatanus	Endog	ABA inhibits germ	present in embryos, CK can overcome	147
A. pseudoplatanus	Endog	↑ loss of neutral inhib from embryo	when strat at 5°C, but not at 20°C	145
A. saccharum	Endog	inhibitors ↓ during strat ↓ germ	p- and o-coummaric, ferulic acids only p-coummaric active	35
A. saccharum	Endog	ABA ↓ during strat	CK ↑ 20-40 d, GA ↑ 40+ d	142
A. saccharum	Endog	98% ABA lost during strat	[CK] and [GA] also change	144

Species	Source of PGR	Effect	Other treatments	Citation
A. tataricum	Endog	ABA ↓	cold strat	89
	Exog	germ ↓ when ABA added	nondormant embryos	
	Endog	ABA ↓, but no germ	warm strat	
Fraxinus americana	Endog	germ with high ABA in embryo and endosperm	excised embryo	38
F. americana	Endog	4X ABA of F. ornus	deeply dormant, require chilling	113
	Exog	ABA ↓ germ	excised embryo, rev by GA, CK	
F. americana	Endog	high ABA	4- to 10-yr-old seeds	116
F. americana	Exog	ABA ↓ germ	antagonistic to zeatin effects	131
F. excelsior	Endog	inhibitors impose dormancy	chilling ↑ stim in embryo	137
F. excelsior	Endog	ABA maintains dormancy	inhibits incorp of uridine and thymidine, antagonistic to GA effect	135
F. ornus	Endog	ABA low	nondormant spp	113
F. lanceolata	Endog	embryo growth with high [ABA]	remove seed coats	81
F. excelsior		ABA ↓	warm or cold strat	
D. AUXINS (IAA, NAA)				
Acer platanoides	Exog	IAA did not induce germ	isolated unchilled embryos	28
A. platanoides	Endog	low IA high IAA	during chilling and 17°C storage in freshly harvested seeds and late in strat	125

Species	Source of PGR	Effect	Other treatments	Citation
Betula varrucosa	Exog	↑ germ capacity		24
F. lanceolata	Endog	IAA, NAA ↑	warm strat	81
F. excelsior		IAA, NAA ↓	cold strat	
E. ETHYLENE				
Cercis canadensis, *Elaeagnus augustifolia*	Exog	↑ germ unstrat seeds also ↑ germ strat seeds	soaked in ethrel solns reduced effect	41

Table 2. Effects of plant growth regulators on germination of forest tree seeds (Gymnosperms)

Species	Source of PGR	Effect	Other treatments	Citation
A. GIBBERELLIN (GA)				
Picea abies	Exog	GA ↑ germ		103
P. abies	Exog	GA soln ↑ germ	germ in soln	114
P. glehnii	Exog	GA ↑ germ	rev BA inhibition, also in 20 yr seeds	107
P. smithiana	Exog	↑ germ	soaked seeds for 48 h	22
P. caribaea	Exog	GA ↑ germ (5-fold)	low vigour seeds, coats removed	134
P. densiflora	Exog	GA ↑ germ	soak several days at 1°C	39
Pinus jeffreyi P. lambertiana P. ponderosa	Endog	GA ↑ as seeds mature no GA activity in coat	embryo, major site of production lower levels than reported for other seeds	58
P. lambertiana	Endog	GA ↑ (20 fold↑)	strat, also ↑ with coat removal	119
P. sylvestris	Exog	germ not affected	water soak, grown on GA-soaked paper	112
P. taeda	Endog	no GA-like high GA-like	unstrat seeds during germination	8
P. taeda	Exog	GA ↑ germ	best when seeds strat 3 weeks	9
P. taeda	Endog	GA-like low, GA-like ↑ no change, GA-like ↑ no GA-like, high GA-like	unchilled dormant seeds, cold strat warm strat, during germ unstrat seeds, during germ	87

Species	Source of PGR	Effect	Other treatments	Citation
Pseudotsuga menziesii	Exog	GA↑ germ rate, not total germ	effective in both light and dark	98
P. menziesii	Endog	GA↑ (7-fold), then GA↓	strat	119
Taxodium distichum	Exog	GA↑ germ	unstrat, slight effect with strat	9
Taxus baccata	Exog	GA↑ germ	rev ABA (exog) imposed dormancy	59
B. CYTOKININ (CK)				
Picea abies	Exog	CK did not affect germ		114
P. abies	Exog	BA did not stim endog CK	BA action not via change in endog CK levels	132
Picea glehnii	Exog	BA↓ germ	reversed by GA_3 and ethrel	107
Pinus lambertiana	Endog	CK↑ 200 fold	strat, also ↑ with coat removal	119
P. taeda	Exog	CK↑ germ	best with 3 wk strat	9
Pseudotsuga menziesii	Endog	CK no change, then slight ↑	strat, then 22°C	119
Taxodium distichum	Exog	no effect on germ	unstrat, or strat	9
C. INHIBITORS (ABA)				
Picea abies Pinus sylvestris	Exog	no effect on germ both treatments >90%	buffers were used, but uncertain if ABA penetrated seeds	50

Species	Source of PGR	Effect	Other treatments	Citation
Pinus koraiensis	Endog	ABA high, then low germ not related to ABA	unstrat seedcoats, then chilling or washing	117
P. lambertiana	Endog	seedcoat [ABA] embryo ABA 2X coat ABA	during soaking or strat ABA not related to dormancy	72
P. lambertiana	Endog	no change, embryo ABA at 25°C, 3-fold ↓ ABA	after 30 d at 5°C but seed still dormant	73
P. pinea	Endog	↓ germ, present in embryo and megagametophyte	probably ABA, also inhibitor similar to phosphon D	27
P. taeda	Endog	inhib ↑, then no inhib	unstrat, then germ	8
P. taeda	Endog	inhibitors↓	cold strat	87
Taxus baccata	Endog Exog	[ABA] ↓, germ ↑ ABA ↓ germ	leached embryos ABA applied to leached embryos	59
T. baccata	Endog	embryo dormancy	leaching in liquid culture	60
D. AUXIN (IAA)				
Pinus lambertiana	Exog	IAA ↓ root growth	14 d embryos with cotyledons	15
P. ponderosa	Exog	↑ elongation hypocotyl	excised embryos	123
P. sylvestris	Endog	IAA ↑ in germ [IAA] 70X [4-Cl-IAA]	2-fold ↑ during 5 d germ mature seeds	36
P. taeda	Exog	↑ elongation hypocotyl	excised embryos, also by fusicoccin	20

Species	Source of PGR	Effect	Other treatments	Citation
P. taeda	Endog	no detectable IAA auxin-like detectable	unstratified seeds after 14 to 56 d strat	87
E. ETHYLENE				
Araucaria hunsteinii	Exog	↓ viability	seeds stored at 30% m.c. and 25°C in 0.01% ethylene and 21% oxygen	126
Picea glehnii	Exog	↑ seed germ	seeds stored for 20 yr	107
Pseudotsuga menziesii	Exog	↑ seed germ inhibition of germ	pure ethylene 12 h before germ high concentration during germ	14
Taxodium distichum	Exog	↑ germ strat and unstrat seeds	soaked in ethrel solutions	41

compared to either GA or ABA (Tables 1, 2). Their activity is more prominent when combined with other promotive agents, such as GAs, light, and ethylene. Cytokinins also are often more effective than GA in counteracting inhibitors (such as ABA) of various GA-sensitive processes. This modifying effect on other hormonal activities led to the consideration of CK as being "permissive" for seed dormancy (52).

Cytokinins are ubiquitous in plants either as free substances or as structural components of tRNA. The many naturally occurring CKs are derivatives of N^6(Δ^2-isopentenyl) adenosine. Zeatin, as a free base and as its ribonucleoside and ribonucleotide, is widely noted as a natural component of seeds (124). Besides the naturally occurring CKs, many synthetic compounds have been tested for activity, of which N^6-benzylaminopurine (BA) is probably the best known.

3.3. Auxins (IAA)

The effect of hormones of the indoleacetic acid (IAA) type on germination has long been in dispute. Numerous workers have investigated the effects of IAA and similar substances on the germination of a variety of seeds, and have obtained conflicting results, stimulation or inhibition being obtained, depending on the concentration of IAA and the type of seed used. However, the most general effect is an absence of response of the seeds to physiological concentrations of IAA (Tables 1, 2). In general, therefore, it must be concluded that IAA can stimulate germination under special conditions, but normally it has little or no effect on most forest tree seeds.

3.4. Abscisic acid (ABA)

Abscisic acid is an endogenous component of dormant and nondormant seeds, as well as an effective inhibitor of germination when supplied exogenously (Tables 1, 2). Evidence for the role of ABA in seed dormancy comes from interactions with GAs and CKs which also affect germination (52). While ABA often counteracts the effects of applied GA in various *in vitro* systems, GA does not always counteract the ABA effect, and frequently a CK is required. Leaching of ABA-inhibited seeds also effectively overcomes inhibition (see Section 4.1.).

4. ROLE OF PLANT GROWTH REGULATORS IN GERMINATION

Although it is conceptually convenient to categorize dormancy types, the real interest (expressed or implied) has concentrated on the control of dormancy. For this purpose, most studies only refer to embryo dormancy and coat-imposed dormancy. This simplified division more closely reflects the fundamental cause of dormancy, which is the inability of the embryonic axis to overcome the constraints acting against it; constraints which reside within the embryo itself (embryo dormancy) or belong to the tissues surrounding it (coat-imposed dormancy).

Whatever the source of constraint, the end result is a failure of the radicle to emerge from the seed, the event usually referred to as germination. Numerous events occur during germination and a block of any one could account for the failure of the embryonic axis to grow. The problem is to discover where the block(s) lie, and consequently, this is where most of the attention in seed physiology research has been focussed (64). From this research a number of concepts have emerged to explain the principles underlying the induction and release of dormancy. These concepts have been built around such observations as (1) the barrier effects of seed coats and permeability changes, (2) the presence of inhibitors and the selective roles of hormones, (3) light and phytochrome effects, and (4) metabolic changes (53).

These concepts and their associated models have been developed largely on evidence obtained from herbaceous crop and weed seeds. If a role is to be established for growth regulators in tree seed germination, it must be demonstrated not only that PGRs are somehow involved in the mechanism(s) that control the induction and release from dormancy, but also that these mechanisms are operative in the seeds of woody plants. The approach taken here will be to review these four concepts of dormancy, citing wherever possible the published evidence relating hormones to the physiology of tree seeds. The rationale for this approach is that the effects of hormones cannot be considered in isolation from the means by which their action is mediated. Discussing the evidence in the context of currently known mechanisms of dormancy release may help us gain further insight into the role of PGRs in the germination of forest tree seeds.

4.1. Barrier effects of seed coats and permeability changes

Seeds can remain viable, yet be unable to germinate, because of the seed coat or other outer structures (endosperm, testa, pericarp) surrounding the embryo (63). Coats may constitute barriers to germination due to physical or chemical characteristics, or they may restrict the permeability to water, gases, or solutes.

Seed imbibition is essential to reactivate stored metabolites and organelles and to initiate the early stages of germination. However, this critical step may be prevented by the presence of impermeable seed coats which restrict the availability of water and gases to inner parts of the seed. Water absorption measured in five southern pines (Pinus echinata, P. elliottii, P. palustris, P. taeda, and P. X sondereggeri) confirmed that seed coats restricted the amount of water moving into the seeds (5). In P. taeda, the most dormant of the species tested, seeds did not exceed 36% moisture content until the coats cracked and germination began. In Acer saccharum, rates of imbibition and water flow measured with an osmometer showed that the testa severely restricted water movement (143). Since a significant portion of the stratification period was required for adequate water uptake, the authors suggested that testa permeability partly controlled germination. In Pinus taeda, seed coats did not limit initial water uptake but did restrict the total amount of water absorbed by the seeds prior to germination (4).

Poor penetration may explain why growth regulators have not been shown to affect germination in many tree seeds, since exogenous application of PGRs is more effective if tissues surrounding the embryo are removed or made more permeable. The impermeability of the seed coat to hormones was cited as a reason for the ineffectiveness of exogenous GA in relieving dormancy of Pinus lambertiana (119). In Pinus albicaulis, the thick coat not only restricted water uptake, but also limited the effectiveness of PGR treatments (96). Difficulty was also experienced in applying GA_3 solutions to seeds of Carya illinoensis, Fraxinus pennsylvanica, Liquidambar stryaciflua, Liriondendron tulipifera, and Quercus falcata (11). Perhaps dissolving the chemicals in organic solvents, as has been done for a variety of agricultural seeds (34,54,68,86,97,118), might improve the success of hormone treatments. The use of organic solvents is suggested with reservation, however, because of the known sensitivity of tree seeds to these agents (3,32,108).

Seed coats also act as physical barriers to the expansion of the embryo. Seeds with coat-imposed dormancy frequently contain nondormant embryos whic germinate quite readily once the coats are removed. However, in intact seeds, the balance between the expansive force of the embryo and the strength of the coat determines whether germination will occur. In some instances, the balance may be tipped in one direction or another by treatments such as stratification. In *Pinus taeda* and four other southern pines, stratification was able to override the physical restriction imposed by the seed coats by enhancing the growth capacity of the embryo (4,5). Carpita *et al.* (19) also studied *Pinus taeda* and the effects of stratification on the growth potential of the embryo. They concluded that growth was due to increased turgor in the embryonic axis which enabled the embryo to overcome the restraint of the megagametophyte and the coat. Although hormones were thought to be important in stratification, no relation between dormancy and hormones was established.

The involvement of hormones in coat-imposed dormancy has been more clearly defined in other studies. Both ABA and seed coats were investigated as the cause of dormancy in *Pinus koraiensis* (117), but neither water permeability nor ABA content in the coat, megagametophyte, or embryo were closely correlated with germination. Fracturing the coat resulted in increased radicle emergence and thus it was concluded that mechanical resistance or gas permeability of the coat was the factor controlling seed dormancy. Seed coats were also of greater significance than hormones in several ash (*Fraxinus*) species since removal of the seed coats allowed embryos to grow in spite of high IAA and ABA contents in the embryo axis (81). On the other hand, hormones may sometimes act synergistically with seed coat removal. Germination of excised *Fraxinus americana* embryos approached 100% if exogenous zeatin was also applied (38). This treatment yielded better results than the usual stratification treatment. *Fagus sylvatica* seeds were induced to germinate by removing pericarps and internal coats, but GA_3 was also required for the treatment to be effective (12).

The emphasis thus far has been on the physical inhibition of germination, but hormones may act more directly in coat-imposed dormancy due to the presence of chemical inhibitors in the seed coats. Several inhibitors, one being similar to ABA, were found in the coats of *Pinus pinea* (62). These inhibitors appeared to regulate germination since washing the seeds for 24 h increased germination to 65% as compared to 25% for controls. The water soluble substances extracted from the *P. pinea* seed coats, when applied exogenously to lettuce and radish seeds, were also able to inhibit germination of other species.

However, inhibition cannot always be attributed to the presence of ABA in the coats. In *Pinus koraiensis*, high ABA levels were found in coats of unstratified seeds (117), but the breaking of dormancy could not be correlated with reduced levels of the inhibitor, even though washing or stratification substantially decreased ABA contents. Similarly, Murphy and Noland (72) showed that ABA declined in *Pinus lambertiana* with leaching, coat removal and various chilling periods. Yet the only treatment that gave appreciable germination was 90 days of stratification (Tables 3, 4).

Plant growth regulators other than ABA may be affected by the presence or absence of the seed coat. In *Pinus lambertiana* seeds, coat removal increased endogenous CK and GA levels in coatless seeds as well as promoting germination (119). The enhancement of these activities by coat removal was not fully explained, but later studies excluded ABA as the causative agent (72).

In certain species the mechanism by which the seed coat exerts its effect may change during the course of germination. ABA present in the testa may

TABLE 3. ABA content of sugar pine seed coats during stratification (72).

Treatment	ABA content (ug/g ODW)
Dry seed coat	3.73 + 0.19 a
24-h soak	1.49 + 0.15 b
30-day stratification	1.04 + 0.11 c
60-day stratification	0.80 + 0.13 cd
90-day stratification	0.58 + 0.09 d

TABLE 4. Germination response of sugar pine seeds to various treatments (72).

Treatment	Germination (%)
Control	5
24-h wash	2
Cracked	2
Seed coat removed	12
90-day stratification	95

maintain dormancy prior to the onset of embryo dormancy. The testa was found to be responsible for dormancy in *Acer pseudoplatanus*, but the mode of action varied with time (146). In early germination the permeability of the testa restricted the outward movement of inhibitors, but later the physical constraint of the covering structures delayed the germination of seeds. Dormancy could be overcome either by removing the testa or chilling the seeds at 5°C. Sometimes chemical inhibition of the coat can be reversed by exogenous growth promotors. Dormancy of *Comptonia perigrina* was found associated with the presence of a chemical inhibitor, possibly ABA, in the testa (29). Soaking scarified seeds in GA_3 for 24 h counteracted this inhibition and resulted in 80% germination.

Seed coats may impose dormancy by excluding light. In one study, the transmittance of natural spectrum flourescent light by upper (convex) and lower (flat) seed coats of Douglas-fir (*Pseudotsuga menziesii*) was measured before and after soaking in water for 3 h (98). Upper coats transmitted about 20% of incident light when dry and about 25% when wet, but transmittance of the lower coats remained about 17% whether the seeds were dry or wet. Drying up to 24 h did not diminish light permeability, indicating that soaking either effected a permanent structural change to the coats, or leached some water-soluble substance. Light exclusion by seed coats has also been observed in other light-requiring seeds. In *Pinus silvestris*, the phytochrome-mediated control of germination was overcome by either puncturing the coats or decoating the seeds (83).

4.2. The presence of inhibitors and selective action of hormones

4.2.1. General remarks

Early studies evaluating the role of PGRs in the control of seed dormancy tended to focus on the effects of one or two classes of hormones, usually GAs and ABA (2). A different perception of hormonal control began to emerge when a third major class of compounds, the CKs, were found to significantly affect seed dormancy (51). A unifying hypothesis was soon proposed in which the interactions between GAs, inhibitors, and CKs in regulating seed germination were explained in terms of primary, preventive and permissive roles, respectively (52). According to this hypothesis, GA has a primary role in promoting seed germination, but only in the absence of inhibitors. When inhibitors are present, CKs are also required to overcome the negative effects of endogenous regulators (Table 5). The hypothesis was based on evidence obtained from a variety of herbaceous seeds, but while application to coniferous seeds has been suggested (99), there does not as yet appear to be any published research relating Khan's hypothesis to seeds of woody

TABLE 5. HORMONAL MECHANISM OF SEED DORMANCY AND GERMINATION (51).

Gibberellin (Primary)	Cytokinin (Permissive)	Inhibitor (Preventive)	
+	+	+	
+	+	-	GERMINATION
+	-	-	
+	-	+	
-	-	-	
-	-	+	DORMANCY
-	+	-	
-	+	+	

plants. By and large, other theories of hormonal regulation have been used to interpret the effects of plant growth regulators on forest tree seeds. For this reason and because the data on hormonal regulation of forest tree seeds is so incomplete, it may be more appropriate for the present to review what is known about those angiosperm genera in which plant growth regulators have been most intensively studied.

Exogenous hormones have been employed in many of the following studies, usually in the expectation that such treatments would overcome the inherent dormancy commonly observed in the seeds of woody plants. Although the application of exogenous PGRs is appropriate in some instances, a more direct approach may prove more suitable for determining the role of PGRs *in vivo*, i.e., by examining endogenous levels of hormones in seeds, and relating these changes to changes in seed dormancy status. Whether changes in endogenous hormones are causally related to the termination of dormancy, or the initiation of germination is an issue which remains to be clarified.

4.2.2. *Acer*

In *Acer saccharum*, the breaking of dormancy appears to consist of an ordered sequence of events that includes progressive changes in endogenous growth hormones (144, 142). Shortly after the start of stratification, there is a dramatic decrease in ABA which continues throughout chilling at 5°C (Figure 1). This is followed, at 20 days, by a peak in CK activity which gradually diminishes until about 40 days (133). At day 40, GA-like substances increase, but then decline with further chilling. This sequence of events indicates that the breaking of dormancy results from an increase in CKs which overcome the effects of endogenous inhibitors, e.g., phenolics such a p-coumaric acid, which act with ABA in maintaining dormancy (35).

Dormancy in *A. pseudoplatanus* and *A. platanoides* also appears to be under the control of endogenous inhibitors (147). ABA levels, which are high in dormant *A. pseudoplatanus* seeds, decrease to low levels by the end of stratification (145). Cytokinins appear to be significant in the breaking of dormancy, and an early peak in CK activity seems to be a common feature of endogenous changes found in *Acer* seeds during chilling (49).

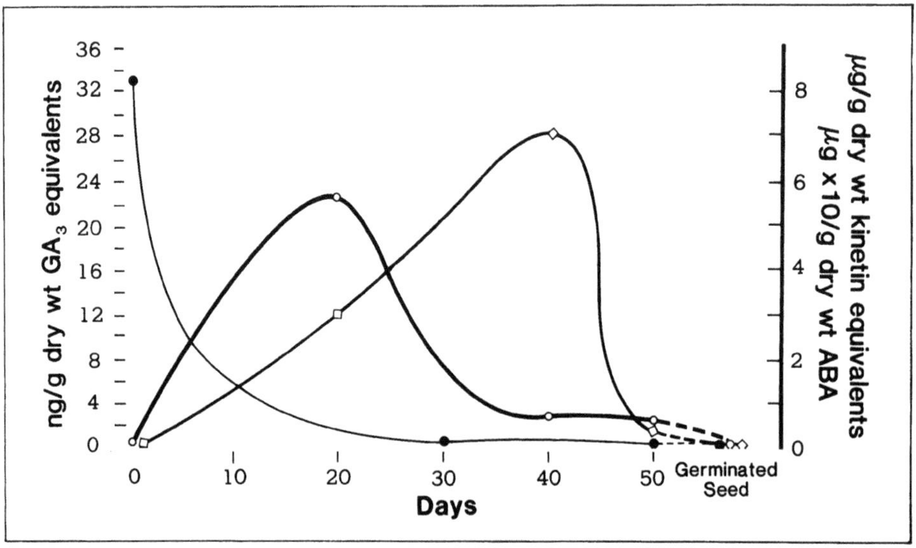

FIGURE 1. The effects of stratification at 5°C on the level of CKs, GAs, and endogenous ABAs. □ Acidic GA-like substances measured by the lettuce hypocotyl bioassay. ○ Butanol-soluble CKs measured by the soybean callus bioassay. ● Endogenous ABA measured by GLC. (144).

Cold temperatures (5°C) are apparently essential to release Acer seeds from dormancy. Only low temperatures were able to overcome dormancy in embryos of A. platanoides, although in later stages CK was also effective (93). Inhibitory activity was lost at 5°C from A. pseudoplatanus embryos, but did not decline at 20°C (145). In A. tataricum, ABA decreased during both warm and cold stratification, but seeds given warm stratification germinated only minimally (89). Embryos did not germinate and endogenous CK levels were low when A. pseudoplatanus seeds were stratified under warm conditions (48).

Although some PGR effects vary between closely related species, the results of numerous studies conducted with exogenous hormones provide an indication of how PGRs may function in in vivo systems. In A. tataricum, ABA inhibited the germination of nondormant seeds (89), whereas GA_3 treatment promoted the germination of excised embryos of the same species (78). GA_3 was not effective, however, in promoting the germination of dormant A. pseudoplatanus seeds, which only responded positively to exogenous CK (147). Exogenous ABA counteracted the effects of endogenous hormones of A. pseudoplatanus, and inhibited radicle and cotyledon growth, usually promoted by CK and GA, respectively (95). As with endogenous systems, exogenous hormones may interact with the temperature conditions given during treatment as seen, for example, in A. saccharum, where exogenous GA and CK increased the growth of embryonic axes during cold (5°C), but not during warm (20°C) stratification (109).

The efficacy of applied PGRs may vary depending upon when in the developmental stage of seeds the treatment is given. Three growth phases could be discerned when exogenous CK and GA were applied during stratification of *A. saccharum* embryos (109). Initially growth regulators had no effect on axis elongation, but this stage was followed by a period of stimulation and, once germination had begun, a period of inhibition. Similar results were obtained for *A. pseudoplatanus* and *A. platanoides* seeds whose sensitivity to CK changed soon after testa removal (30). It has recently been argued that it is the change in sensitivity to plant growth regulators, rather than the absolute amounts of hormones, which govern plant growth and development (129, 130). In view of this suggestion, the results reported above for *Acer* may have some interesting theoretical implications.

4.2.3. *Fraxinus*

ABA is clearly involved in the dormancy of *Fraxinus* seeds. High ABA levels were found in both embryo and endosperm tissues of dormant *F. americana* seeds (38). Levels of the inhibitor appeared to be directly related to the degree of dormancy, since deeply dormant *F. americana* seeds had four times the amount of ABA as nondormant *F. ornus* seeds (113). ABA was present in all tissues, but the highest concentrations were found in the seed and the pericarp. However, high levels of endogenous hormones may not always have an identifiable physiological significance. Additional experiments with applied ABA indicated a regulatory role for ABA in the seed, but no apparent function for ABA in the pericarp. Poor correspondence between hormone levels and inhibition was also seen in *F. lanceolata* and *F. excelsior*, since coat removal allowed stratified embryos to germinate in spite of high ABA and IAA contents (81). The authors explained this anomaly by speculating that freeing the emryos from the surrounding integuments activated enzymes that decreased ABA and IAA contents of embryo tissues. Some validity for a direct relationship between ABA and embryo inhibition was shown in another study in which exogenous ABA prevented germination when it was applied to nondormant embryos of *F. americana* and *F. ornus* (113).

Since chilling is the most effective way to overcome dormancy in *Fraxinus* seeds, several studies have examined the relationship between hormones and low temperatures. Chilling of *F. americana* decreased ABA levels 37% in the pericarp and 68% in the seed, so that after cold treatment ABA concentrations in *F. americana* seeds were as low as levels found in nondormant *F. ornus* seeds (113). *F. excelsior* embryos produce stimulating substances (tentatively identified as indoleacetonitrile) during chilling which are able to overcome the dormancy imposed by inhibitors present in the embryo and the endosperm (137). Despite the apparent connection between ABA levels and breaking of dormancy in *F. americana*, Sondheimer *et al.* (113) did not feel that reductions in inhibitor levels during stratification were a general control mechanism in all seeds which require chilling for dormancy release.

Germination of *Fraxinus* seeds appears to be promoted by both CK and GA. Treating excised embryos of *F. americana* with zeatin resulted in 99% germination, compared to 37% for seeds which had been chilled for 30 days (38), but the effect could be reversed by ABA (131). Relative to CK, gibberellin may be a more effective phytohormone for exogenous application to *Fraxinus* seeds, since GA was able to enhance germination of intact and unstratified seeds of *F. americana* (65), and *F. pennsylvanica* (112).

Some indication of the complexities of dormancy control has been demonstrated in studies using exogenous GA. For example, F. excelsior seeds may require both warm stratification (20°C) to overcome morphological dormancy (immaturity) and cold stratification (5°C) to overcome physiological dormancy. Nevertheless, Wcislinska (141) observed that exogenous GA had the same effect as warm stratification in removing the morphological immaturity of the embryos. Similarly, Villiers and Wareing (138) found both exogenous GA_3 and leaching effective in promoting germination of dormant embryos.

In contrast, exogenous ABA maintained both leached and chilled embryos in a dormant state (135). However, these inhibitory effects of exogenous ABA could be reversed if F. excelsior embryos were treated with a combination of exogenous GA and CK (113).

4.3. Light and phytochrome effects

Light is required for the germination of many forest tree seeds (44). Control by light is evidently mediated by the phytochrome system, as a number of tree seeds have been shown to respond to alternating red and far-red light in a reversible manner (13,67,127,128). As with other phytochrome-controlled processes, phytochrome is probably not directly involved in dormancy-breaking reactions, but acts, due to the photoreversible nature of the pigment, as an on/off mechanism to inititate a series of events that ultimately result in germination.

Although not directly responsible for dormancy release, it is evident that phytochrome is intimately associated with the dormancy process. However, the number of reported interactions between light and stratification, and light and plant growth regulators, attest to the fact that such a complex phenomenon cannot be attributed to a single factor such as light. For example, even though stratification is widely recognized as the most consistently effective method for breaking dormancy in tree seeds, light can in many instances partially or completely replace the chilling requirement (1,66,115). Conversely, seeds which require light will germinate in the dark if they are given sufficient exposure to chilling temperatures. Germination of unstratified Pinus ponderosa seeds was promoted by red light and inhibited by far-red light (42), but seeds lost all light sensitivity after stratification for 21 days. In Tsuga canadensis, however, stratification did not render seeds less responsive, but instead enhanced the effects of red light treatment (26).

Similar interactions have been shown to occur between plant growth regulators and light. Skotodormancy was induced in Paulownia tomentosa by imbibing the seeds for 30 days in darkness, however, this state could be broken by exposing the seeds to either 6 h red light or exogenous GA_3 (40). Exogenous application of CK retarded the germination induced by red light, but the hormone had no effect on germination promoted by GA_3. Since CK affected the light-induced response, but not the response to GA_3, it was concluded that skotodormancy was related to separate reactions activated by phytochrome. Subsequent steps in germination were likely under hormonal control. Almost the same conclusion was reached by Coffman (26) who felt that GA was not directly involved in germination of Tsuga canadensis which is mediated by red light. Gibberellic acid, alone or in combination with kinetin, inhibited germination, whether or not the seeds had received red light exposure. In Pseudotsuga menziesii, light accelerated germination of unstratified seeds, and GA increased rates in both the light and in the dark (98). However, it was concluded in this case also that GA did not interact with light per se, but stimulated germination in some other way.

The significant changes induced by light, and the number of interactions between light and other dormancy releasing factors question the manner in which light is able to effect these changes. It has been suggested, in the case of stratification, that interactions with light result from the effects of cold temperatures on phytochrome. Seeds of Betula papyrifera can be induced to germinate by chilling at 3°C or by red light treatment (6). Cold temperatures enhanced the effect of red light treatments, and red light given during stratification prevented declines in germination due to prolonged chilling. From these results and from indirect evidence of temperature effects on membranes, the authors suggested that cold temperatures either increased the sensitivity of Pfr receptor sites, or acted to retain more of the phytochrome in the Pfr (active) form.

Light may promote germination of tree seeds by enhancing the synthesis of endogenous plant growth regulators such as GA_4 (10). The importance of light quality and the involvement of phytochrome on germination and hormone levels was demonstrated in Pinus silvestris by Kopcewicz (55). Continuous irradiation of seeds with red, orange, or blue light stimulated germination, increased GA contents, and lowered inhibitor levels. Irradiation with green light produced none of the above effects. Further studies with the same species showed that either stratification or red light was effective in promoting germination and simultaneous enhancement of endogenous gibberellin levels (56,57). Similar increases in endogenous CK and GA levels were found in Pseudotsuga menziesii and Pinus lambertiana seeds after stratification (119), and in Picea sitchensis seeds after red light treatment (120).

4.4. Metabolic changes

The essential systems of most seeds are inactivated by the time they reach maturity. Some of these systems, such as soluble enzymes or mitochondria, are easily reactivated by water during imbibition, but others, such as ribosomes or long-lived mRNA, need to be freed from protein before they can become functional in protein synthesis (25).

The germination process can be divided into three distinct, but interacting phases: reactivation of preexisting systems during imbibition; synthesis of enzymes and organelles for degradation of reserves; and synthesis of new cellular components (25).

The major event of the first phase is reactivation of conserved systems to start cellular metabolism such as protein synthesis, solute and ion transport, and respiration. Seed germination requires energy, and adenosine triphosphte (ATP) is regarded as the primary energy source for the process. However, dry seeds do not accumulate ATP, and thus its production at the early stage of germination is essential for physiological function (88). ATP is built up after imbibition (82) through glycolysis, fatty acid oxidation, respiration, and synthesis of adenosine diphosphate (ADP). Substrates for respiration and protein synthesis must also be provided. Sugars and fatty acids (84) are common substrates for respiration which are released from starch and triglycerides, respectively, by the action of pre-existing hydrolytic enzymes. Storage proteins provide a source of amino acids and reduced nitrogen.

The second, or degradation, phase occurs primarily in seed storage organs (cotyledons, endosperm or megagametophyte). The activity of mitochondria, glyoxysomes, enzymes, and substrates increases during or after imbibition, reaches a peak at about 50% exhaustion of reserves, and then declines until reserves are completely depleted. These reactions produce raw materials, reduced pyridine nucleotides, and energy supplies which are transported from the storage organs to the embryo.

Synthesis of new molecules predominates in the third phase of germination; however, activity is dependent upon environmental conditions as well as upon the products generated by the previous two phases. The net result is the production of energy and substrates which provide the enzymatic and structural components essential for the growth and emergence of the embryo.

What is the evidence linking PGRs with the various phases of germination? By what mechanisms do PGRs influence seed metabolism during germination? Protein and lipid storage bodies are clearly evident in the dry seeds of forest trees (21,69,70,110,111,136). The mobilization of such reserves upon imbibition depends upon specific enzymes which hydrolyze stored molecules (82,101,102), and catalyze essential reactions in energy-generating cycles (31,72). Demonstration of PGR involvement in the activation of key enzymes would indicate a role for PGRs during the initial dormancy-breaking stages of germination.

In angiosperms, both endosperm and cotyledons serve a nutritive function, but in gymnosperms, the megagametophyte acts as the reservoir of proteins and lipids needed to sustain the early growth of the embryo. However, lipids are not transported readily to the embryo and must first be transformed into sugars via the glyoxylate cycle. Increases in the specific activity of a key enzyme of the glyoxylate cycle, isocitrate lyase (ICL), can be correlated with corresponding decreases in the lipid content of the megagametophyte. In *Pinus lambertiana*, ICL activity increased in both the embryo and megagametophyte during stratification at 5°C (82). During germination, ICL decreased in the embryos, but rose continuously in megagametophyte tissue. This pattern of mobilization reflects that seen in *Pinus sylvestris* and *Picea abies*, in which the megagametophyte supplies the embryo after the embryo has used up the bulk of its own supplies (110, 111).

The dependence of the embryo on megagametophyte reserves suggests some coordination between mobilization in the embryo and the megagametophyte, and there is some evidence that PGRs are involved. ICL increased in the megagametophyte of *Pinus ponderosa* seeds during stratification and germination (8), but activity was greatly diminished when embryos were removed. Embryo extracts contained a factor which enhanced ICL activity in isolated megagametophytes, but the factor could not be replaced by IAA, GA_3, or BA. Freshly harvested hazel (*Coryllus avellana*) seeds are dormant at 20°C, but can be induced to germinate by chilling or by application of exogenous GA (91). Exogenous GA increases ICL activity in the cotyledons, mobilizing the stored lipids which are essential for germination. Nyman (85) also found that amylase activity of *Pinus sylvestris* megagametophytes was dependent on the presence of the embryo, and that amyolytic activity could not be restored by treating excised megagametophytes with GA_3. There appears to be an important interaction between the embryo and nutritive seed tissues, and although the nature of the substance produced by the embryo has not been ascertained, a role for PGR should not yet be excluded. Responses to GA treatment are dependent upon experimental conditions, and can vary considerably with the type of GA and the test environment (100).

On the premise that the breaking of dormancy and onset of germination may involve the synthesis of specific proteins and enzymes (71), it has been of interest to determine if applied PGRs can modulate transcription and/or translation and induce quantitative or qualitative shifts in protein synthesis. Several direct biochemical studies have been made on the effects of GA on protein and nucleic acid synthesis during germination.

Using autoradiography, Villiers (135) showed that ABA maintained dormancy in Fraxinus excelsior by its effects on specific types of mRNAs, and therefore the formation of specific proteins. ABA inhibited the incorporation of (^{3}H)uridine and (^{3}H)thymidine, while GA_3, which promoted the formation of both RNA and DNA, reversed the ABA effect on germination.

Application of GA_3 released Coryllus avellana seeds from dormancy and stimulated DNA synthesis (45, 46, 94). It was suggested that GA-induced germination in Corylus is due to the derepression of certain specific genes, allowing the production of mRNAs for proteins which are essential to germination (46).

Jarvis and Shannon (47) subsequently examined changes in polyadenyleted RNA (poly $(A)^+$RNA, presumptive mRNA species) metabolism during dry storage and dormancy release, and related them to concommitant changes in GA_3 and ABA. During dry storage, ABA and the incorporation of $^{32}PO_4^{3-}$ in poly $(A)^+$RNA declined, but poly $(A)^+$RNA synthesis increased again when whole seeds were imbibed. Involvement of PGRs was indicated, since germination and poly $(A)^+$RNA synthesis was promoted by GA_3, but inhibited by ABA.

Continuing this work, Shannon and Jarvis (106) studied poly $(A)^+$RNA synthesis in dormant and non-dormant cotyledons. They found that synthesis was promoted by cotyledon excision and testa removal, as well as by GA_3. ABA, thought to reside in the coat, was identified as the inhibitor of poly $(A)^+$RNA synthesis in the cotyledons. The authors concluded that there are two independent sites for GA action, one in the embryonic axes and another in the cotyledons. Hormonal interactions which exist between the two sites are initiated by stratification, and are responsible for poly$(A)^+$RNA synthesis.

Examination of mRNA in dormant and non-dormant embryonic axes showed that GA_3 and cotyledon excision changed the size distribution of poly $(A)^+$RNA and type of polypeptides that were synthesized during stratification (105). The authors concluded that the breaking of dormancy is associated with the transcription of novel mRNAs specific for germination, and since GA_3 and ABA were shown to significantly affect both RNA synthesis and germination, these PGRs must be involved in the regulation of Coryllus avellana seed dormancy.

The work just cited on Coryllus avellana is a fraction of similar research that has been conducted on these seeds, but it provides a good example of the benefits to be derived from concentrating on a single species. Nothing of comparable complexity or magnitude has been attempted for coniferous seeds, a great disadvantage in understanding the role of PGRs in germination of gymnosperms. Generally it is not possible to extrapolate results obtained from angiosperms because of basic biological differences between these two types of seed-bearing plants. In gymnosperms, the megagametophyte is the source of stored reserves, but, in contrast to the angiosperms, no physical connection exists between the storage body and the embryo, therefore nutrients and substances such as PGRs are transported only by diffusion. In angiosperms, endosperm and cotyledon tissues are triploid (3N) and diploid (2N), respectively, whereas the megagametophyte is haploid (1N). Thus, the nutritive tissue in conifers, derived exclusively from the female gamete, has no genetic contribution from the male parent. Although the significance of this and other traits specific to gymnosperms are not yet known, undoubtedly the role and mode of action of PGRs will be affected by these differences in physiology. For example, extremely high levels of some IAA derivatives that are found in angiosperm seeds do not occur in Pinus silvestris (36). The low endogenous level of 4-Cl-IAA and the complete absence of 4-Cl-IAA-Me in P. silvestris may indicate that in gymnosperms these PGRs are metabolized differently or perform other physiological roles.

In recent years there has been increasing interest in the involvement of membranes in coat-imposed dormancy (7). The evidence linking dormancy with cell membranes comes primarily from the effects of temperature on germination and dormancy release. In seeds affected by high temperatures, dormancy is not gradually expressed as the temperature rises but develops abruptly when a critical temperature is exceeded. The sharp discontinuity in the response to temperature indicates that cell membranes are involved, since they are known to undergo sharp transitions in their physical state at particular temperatures.

Some support has been found for the effects of temperature on membranes of Pinus lambertiana (74). When excised embryos were incubated above 15 to 20°C, they exhibited significantly higher rates of imbibition and solute leakage. The inability of dormant P. lambertiana seeds to germinate at temperatures above 17°C was thought to result from these initial temperature effects on membrane properties (75). The possibility of membrane regulation of dormancy is significant because of the demonstrated effects of plant hormones on cell membranes. This connection, although intriguing, has yet to be made for the seeds of woody plants.

It is pertinent in regard to cell membranes to note the ability of ethanol, ethyl ether and chloroform to stimulate dormant seeds to germinate. These compounds are classed as anaesthetics whose physiological effects may be due to their interaction with cell membranes (121). If the compounds interfere with the membranes of seeds and thereby break dormancy it would suggest that the dormant condition depends upon the state of those membranes. To test whether germination might be inceased through alcohol effects on membranes, ethanol was applied to Pinus halepensis seeds (90). In the light, there were no differences between ethanol-treated and untreated seeds, but treated seeds germinated better than untreated seeds in the dark. It was speculated that the solvent may act upon specific membrane sites, activating mechanisms associated with embryo expansion. With the seed layer restriction overcome, dark germination could proceed at the same rate as in the light.

5. CONCLUSIONS

The germination of tree seeds appears to be controlled by a variety of external and internal factors. PGRs figure prominently among these factors, but the range of mechanisms by which PGRs control is mediated may vary considerably--from physical to metabolic. Although significant support exists for PGR involvement in the regulation of angiosperm seed dormancy, the evidence for gymnosperms is still inconclusive. This is because our present understanding of the control of germination is unsatisfactory. Our knowledge of dormancy regulation--by PGRs or other means--in tree seeds is superficial and fragmented. Seed dormancy studies tend to be scattered among a limited number of species, and only rarely is any one species investigated extensively. Coordination between various studies is difficult because in some cases investigations have been confined to entire seeds, while in others, only to parts of seeds. Adding to the confusion is the extrapolation from studies of exogenous PGRs to experiments in which endogenous PGR levels are correlated with dormancy-breaking events. Because of the uncertainty of PGR permeability, or the extent of their metabolism, it is impossible to deduce from exogenous PGR experiments the effective concentrations of PGRs at their site of action.

The breaking of dormancy and the initiation of germination are two separate and distinct processes, yet it is apparent from reading the literature that this somewhat obvious fact is often overlooked. Researchers must concentrate on the very earliest stages of germination, i.e., during dormancy release, if they are to learn how PGRs regulate tree seed dormancy. Little attention has been given as well to seed maturation, but in many tree seeds the induction of dormancy takes place during the final stages of seed development (33), and thus the investigation of PGRs levels during this period should be revealing.

The relatively few tree seeds in which PGR research has been reported indicates the necessity for increasing the number of species which are selected for study. The lack of comparative work between gymnosperm genera is especially noted. However, this does not eliminate the need for intensive, in-depth studies of individual species. The detailed studies of Acer, Coryllus, and Fraxinus provide examples of the desirability of cooperative work on the same genera.

Perhaps the most promising development in PGR research has been the critical reevaluation in recent years of the theoretical assumptions underlying PGR control of plant growth and development (37,129,130). According to Trewavas (129), traditional hormone theory is lacking because it assumes that growth substances are limiting, and therefore, that changes in PGR concentrations regulate plant development. He argues that since development is a continuous process in plants, cells and tissues are continually acquiring and losing their capability to respond to growth substances. It is the acquisition and loss of sensitivity, and not the concentration of PGRs, that determine plant response to PGRs. Thus, in terms of understanding development in the intact plant, tissue sensitivity to PGRs has far more relevance than PGR levels.

To appreciate how PGRs control plant development, Trewavas (130) likened metabolic systems to interconnected networks. Enzymes and proteins form the basic constituents of the metabolic "network" and are linked in a variety of ways. The critical property of the network is that external or internal effects on one or more constituents do not take place in isolation, but instead are experienced via linkages throughout the entire network. The model emphasizes the importance of context, since the effects of exogenous or endogenous PGRs are not absolute, but are relative to the internal and external environment of the cell.

As a result of recent advancements in theory and technology, a favourable climate seems to be emerging for PGR research. Growth substance sensitivity is gaining recognition as a useful and valid concept (37). Evidence of its increased use and acceptance within the scientific community is seen in the fact that Firn has encouraged experimental testing of the model and called for stricter definitions of terms.

Substantial improvements are also evident in modern methods of PGR analysis. Relative to older bioassay techniques, new physiochemical methods provide greater specificity and sensitivity, less variability, and shorter analysis times (16). The use of more sensitive methods is already resulting in the identification of PGRs not previously detected in forest tree seeds (36). Particularly promising of the newer methods are immunoassays (61), which can detect very low PGR concentrations in relatively impure solutions. Immunoassays are also economical, rapid, and do not require a highly specialized, heavily equipped laboratory (148). These improvements in PGR technology, together with a more holistic concept of PGR action in biological systems, signal a revitalized approach to the investigation of the role of PGRs in plant growth and development.

REFERENCES

1. Adkins CR, Hinesley LE, Blazich FA (1984) Role of stratification, temperature, and light in Fraser fir germination. Can J For Res 14(1):88-93
2. Amen RD (1968) A model of seed dormancy. Bot Rev 34:1-31
3. Barnett JP (1971) Flotation in ethanol reduces storability of southern pine seeds. For Sci 17(1):50-51
4. Barnett JP (1972) Seedcoat influences dormancy of loblolly pine seeds. Can J For Res 2:7-10
5. Barnett JP (1976) Delayed germination of southern pine seeds related to seed coat constraint. Can J For Res 6:504-510
6. Bevington JM and Hoyle MC (1981) Phytochrome action during prechilling induced gemination of Betula papyrifera Marsh. Plant Physiol 67:705-710
7. Bewley JD and Black M (1982) Physiology and biochemistry of seeds in relation to germination. Viability, dormancy, and environmental control. Vol. 2. New York: Springer-Verlag, 375 p
8. Bilderback, DE (1974) The regulatory role of the embryo on the development of isocitrate lyase activity during germination of ponderosa pine seeds. Physiol Plant 31:200-203
9. Biswas PK, Bonamy PA and Paul KB (1972) Germination promotion of loblolly pine and bald cypress seeds by stratification and chemical treatments. Physiol Plant 27:71-76
10. Bogatek R, Podstolkski A, Ostaszewska A and Lewak S (1976) Phloridzin transformation and accumulation during the stratification of apple seeds and the culture of isolated embryos. Biologia Plantarum 18:241-250
11. Bonner FT (1976) Effects of gibberellin on germination of forest tree seeds with shallow dormancy. Proc 2nd Intl Symp IUFRO S2.01.06, pp 21-32
12. Bonnet-Masimbert M and Müller C (1976) Mise au point d'un test rapide de germination des faînes (Fagus sylvatica). Can J For Res 6(3):281-286
13. Bormann BT (1983) Ecological implications of phytochrome-mediated seed germination in red alder. For Sci 29(4):734-738
14. Borno C and Taylor IEP (1975) The effect of high concentrations of ethylene on seed germination of Douglas-fir (Pseudotsuga menziesii (Mirb.) Franco). Can J For Res 5(3):419-423
15. Brown CL and Gifford EM Jr (1958) The relation of the cotyledons to root development of pine embryos grown in vitro. Plant Physiol 33:57-64
16. Brenner ML (1981) Modern methods for plant growth substance analysis. Annu Rev Plant Physiol 32:511-538
17. Burns RM (1967) Gibberellic acid stimulates germination of sweetgum seed. For Sci 13(4):438-439
18. Cardemil L and Reinero A (1982) Changes of Araucaria araucana seed reserves during germination and early seedling growth. Can J Bot 60(9):1629-1638
19. Carpita NC, Skaria A, Barnett JP and Dunlap JR (1983) Cold stratification and growth of radicles of loblolly pine (Pinus taeda) embryos. Physiol Plant 59:601-606
20. Carpita NC and Tarmann KM (1982) Promotion of hypocotyl elongation in loblolly pine (Pinus taeda L.) by indole-3-acetic acid. Physiol Plant 55:149-154

21. Cecich RA and Horner Jr HT (1977) An ultrastructrual and microspectrophotometric study of the shoot apex during the initiation of the first leaf in germinating Pinus banksiana. Amer J Bot 64(2):207-222
22. Chandra JP and Chauhan PS (1976) Note on germination of spruce seeds with gibberellic acid. The Indian Forester 102(1):721-725
23. Chen SC and Varner JE (1973) Hormones and seed dormancy. Seed Sci Technol 1:325-338
24. Chernobrovkina NP, Kefeli VI and Ivanova RP (1975) Dormancy, growth, and their natural regulators in buds and seeds of Karelian birch. Soviet Plant Physiology 22(5):887-893
25. Ching TM (1973) Biochemical aspects of seed vigor. Seed Sci Technol 1(1):73-88
26. Coffman MS (1975) Tsuga canadensis (L) Carr. germination stimulated by red light while inhibited by gibberellin and kinetin. Research Note No 13, Ford Forestry Center, Michigan Technological University, 12 p
27. Corvillon E and Martínez-Honduvilla CJ (1980) Germination inhibitors in embryos and endosperms from Pinus pinea seeds. Ital J Biochem 29(6):405-411
28. Davies HV (1977) Seed dormancy in Acer platanoides L. PhD thesis, University of Bristol
29. del Tredici P and Torrey JG (1976) On the germination of seeds of Comptonia peregrina the sweet fern. Bot Gaz 137(3):262-268
30. Dungey NO, Brown G and Pinfield NJ (1980) The involvement of cytokinins in the removal of testa-induced dormancy in Acer seed. Z Pflanzenphysiol 96:233-237
31. Durzan DJ (1983) Metabolism of tritiated water during imbibition and germination of jack pine seeds. Can J For Res 13:1204-1218
32. Edwards DGW (1980a) Tree seed research, Pacific Forest Research Centre, B.C., 1977-1979. Proc 17th Mtg Can Tree Improvement Assoc, Part I, pp 231-236
33. Edwards DGW (1980b) Maturity and quality of tree seeds--a state-of-the-art review. Seed Sci Technol 8:625-657
34. Eldan M, Mayer AM and Poljakoff-Mayber A (1974) Permeation of dry lettuce seeds with acetic anhydride and with amino acids using dichloromethane. Seed Sci Technol 2:317-322
35. Enu-Kwesi L and Dumbroff EB (1980) Changes in phenolic inhibitors in seeds of Acer saccharum during stratification. J Exp Bot 31:425-436
36. Ernstsen A and Sandberg G (1986) Identification of 4-chloroindole-3-acetic acid and indole-3-aldehyde in seeds of Pinus sylvestris. Physiol Plant 68:511-518
37. Firn RD (1986) Growth substance sensitivity: The need for clearer ideas, precise terms and purposeful expeiments (Review). Physiol Plant 67(2):267-272
38. Gendel SM, Fosket DE and Miksche JP (1977) Increasing white ash seed germination by embryo dissection. USDA Forest Service Research Note, North Central Forest Experiment Station No. NC-220. Developmental and Cell Biology Dep, Sch of Biol Sciences, Univ of Calif, Irvine, Calif USA, 3p
39. Goo M and Tusihasi H (1958) Hastening the germination of Pinus densiflora seeds by gibberellin. J Jap Forestry Soc 40:509-511
40. Grubisic D, Neskowic M and Konjević R (1985) Changes in light sensitivity of Paulownia tomentosa and P. fortunei seeds. Plant Sci 39:13-16

41. Hamilton DF (1972) Ethrel for breaking dormancy in seeds of woody plants. Int Plant Prop Soc Proc 22:368-373
42. Harrington M (1977) Response of ponderosa pine seeds to light. Forest Service USDA, INT-220, 8 p
43. Hedden P, MacMillan J and Phinney BO (1978) The metabolism of the gibberellins. Ann Rev Plant Physiol 29:149-192
44. International Seed Testing Association (1985) International rules for seed testing--annexes 1985. Seed Sci Technol 13(2):356-513
45. Jarvis BC, Frankland B and Cherry JH (1968a) Increased DNA template and RNA polymerase associated with the breaking of seed dormancy. Plant Physiol 43:1734-1736
46. Jarvis BC, Frankland B and Cherry JH (1968b) Increased nucleic-acid synthesis in relation to the breaking of dormancy of hazel seed by gibberellic acid. Planta 83:257-266
47. Jarvis BC and Shannon PRM (1981) Changes in poly(A) RNA metabolism in relation to storage and dormancy-breaking of hazel seed. New Phytologist 88(1):31-40
48. Julin-Tegelman Å and Pinfield N (1982) Changes in the level of endogenous cytokinin-like substances in Acer pseudoplatanus embryos during stratification and germination. Physiol Plant 54:318-322
49. Julin-Tegelman Å and Pinfield N (1983) Changes in the level of endogenous cytokinin-like substances during cold-induced germination of Acer platanoides L. seeds. Z Pflanzenphysiol 110:89-95
50. Kamra SK (1985) Germination studies on seeds of Pinus silvestris and Picea abies treated with abscisic or cinnamic acids. Indian J Plant Sci 3(1):1-8
51. Khan AA (1971) Cytokinins: Permissive role in seed germination. Science 171:853-859
52. Khan AA (1975) Primary preventive and permissive roles of hormones in plant systems. Bot Rev 41:391-420
53. Khan AA (1977) Seed dormancy: changing concepts and theories. In: The Physiology and Biochemistry of Seed Dormancy and Germination. Khan AA, ed. Amsterdam: North-Holland Publishing Company, pp 29-50
54. Khan AA, Tao KL and Roe CH (1973) Applications of chemicals in organic solvents to dry seeds. Plant Physiol 52:79-81
55. Kopcewicz J (1972) Effect of coloured light irradiation on germination and endogenous growth regulator content in Scots pine (Pinus silvestris L.). Bull Acad Polon Sci, Ser Sci Biol 20 (6):419-424
56. Kopcewicz J and Porazinski Z (1973a) Effect of red light irradiation on metabolism of free and bound gibberellins in Scots pine (Pinus sylvestris). Bull de l'Academie Polonaise des Sciences 221:383-387
57. Kopcewicz J and Porazinski Z (1973b) Influence of low temperature on germnation and endogenous growth regulator contents in Scots pine seeds. Acta Societatis Botanicorum Poloniae 42:233-240
58. Krugman SL (1967) A gibberellin-like substance in immature pine seed. For Sci 13(1):29-37
59. LePage-Degivry MT (1973) Influence of abscisic acid on the germination of the embryos of Taxus baccata L. cultured in vitro. Z Pflanzenphysiol 70:406-413
60. LePage-Degivry MT and Garello G (1973) La dormance embryonnaire chez Taxus baccata: Influence de la composition du milieu liquide sur l'induction de la germination. Physiol Plant 29:204-207

61. Maldiney R, Leroux B, Sabbagh I, Sotta B, Sossoumtzov L and Miginiac E (1986) A biotin-avidin-based enzyme immunoassasy to quantify three phytohormones: Auxin, abscisic acid and zeatin-riboside. J Immunol Methods 90(2):151-158
62. Martinez-Honduvilla CJ and Santos-Ruiz A (1978) Germination inhibitors in the pine seed coat. Planta 141(2):141-144
63. Mayer AM and Poljakoff-Mayber A (1975) The germination of seeds. 2nd ed. New York: Pergamon, 192 p
64. Mayer AM and Shain Y (1974) Control of seed germination. Annu Rev Plant Physiol 25:167-193
65. McBride JR and Dickson R (1972) Growth inhibitor reversed, gibberellic, citric acids and stratification enhance white ash germination. Tree Planters' Notes 23(3):1-2
66. McLemore BF (1971) Light requirements for germination of loblolly pine seeds. For Sci 17:285-286
67. McLemore BF and Hansbrough T (1970) Influence of light on germination of Pinus palustris seeds. Physiol Plant 23:1-10
68. Meyer H and Mayer AM (1971) Permeation of dry seeds with chemicals: use of dichloromethane. Science 171:583-584
69. Mia AJ and Durzan DJ (1974) Cytochemical and subcellular organization of the shoot apical meristem of dry and germinating jack pine embryos. Can J For Res 4(1):39-54
70. Mia AJ and Durzan DJ (1977) Cytochemical and subcellular organization of root apical meristems of dry and germinating jack pine embryos. Can J For Res 7(2):263-276
71. Murphy JB, Murphy JB and Noland TL (1980) Activity of L-phenyalanine ammonia-lyase during stratification and germination of sugar pine seeds. Z Pflanzenphysiol 97(2):161-169
72. Murphy JB and Noland TL (1981a) Changes in phenolic acids and abscisic acid in sugar pine seed coats during stratification. Physiol Plant 52:370-374
73. Murphy JB and Noland TL (1981b) Changes in phenolic acids and abscisic acid in sugar pine seed embryos and megagametophytes during stratification. Physiol Plant 52:375-379
74. Murphy JB and Noland TL (1982a) Temperature effects on seed imbibition and leakage mediated by viscosity and membranes. Plant Physiol 69(2):428-431
75. Murphy JB and Noland TL (1982b) Temperature effects on seed imbibition and leakage mediated by viscosity and membranes. Plant Physiol 70:1410-1412
76. Murray EW and Adams RE (1980) Embryonic control of isocitrate lyase activity in the megagametophyte of ponderosa pine seeds. Physiol Plant 49(1):21-26
77. Nagy M (1980) Dormancy in fruits of Tilia platyphyllos Scop. IV. Changes in the endogenous gibberellin content during stratification. Acta Agronimica Academiae Scientiarum Hungaricae 29:1-11
78. Nikolaeva MG (1969) Physiology of deep dormancy in seeds. Acad Sci USSR. Trans from the Russian by Israel Prog for Sci Translation, Jerusalem
79. Nikolaeva MG (1977) Factors controlling the seed dormancy pattern. In: The Physiology and Biochemistry of Seed Dormancy and Germination. Khan AA ed. Amsterdam: North-Holland Publishing Company, pp 51-74

80. Nikolaeva MG and Judin VG (1963) The effect of gibberellin on the germination of tree seeds. Dokl Akad Nauk SSSR 150(3):686-689
81. Nikolaeva MG and Vorob'eva NS (1979) The role of abscisic acid and indolic compounds in dormancy of the seeds of ash species. Soviet Plant Physiology 26:105-113
82. Noland TL and Murphy JB (1984) Changes in isocitrate lyase activity and ATP content during stratification and germination of sugar pine seeds. Seed Sci Technol 12:777-787
83. Nyman B (1963) Studies on the germination in seeds of Scots pine (Pinus silvestris L.) with special reference to the light factor. Studia Forestalis Suecica Nr 2
84. Nyman B (1966) Studies on the fat metabolism of light- and dark-germinated seeds of Scots pine (Pinus silvestris L.). Physiol Plant 19:63-75
85. Nyman B (1971) Light seed coat and gibberellic acid in relation to the amylase activity in germinating Scots pine seeds (Pinus silvestris). Physiol Plant 25:112-117
86. Palevitch D and Thomas TH (1974) Thermo-dormancy release of celery seed by gibberellins, 6-benzylaminopurine, and ethephon applied in organic solvent to dry seeds. J Exp Bot 25(88):981-986
87. Paul KB, Patel CS and Biswas PK (1973) Changes in the endogenous growth regulators in loblolly pine seeds during the process of stratification and germination. Physiol Plant 28:530-534
88. Perl M (1986) ATP synthesis and utilization in the early stage of seed germination in relation to seed dormancy and quality. Physiol Plant 66:177-182
89. Petrova VN and Nikolaeva MG (1974) The role of native abscisic acid in dormancy of the seeds of tartar maple. Fiziol Rast 21:343-350
90. Petruzzelli L (1984) The effect of ethanol on germination of aleppo pine seeds (Pinus halepensis Mill). Biol Plant 26(3):235-238
91. Pinfield NJ (1968) The promotion of isocitrate lyase activity in hazel cotyledons by exogenous giberellin. Planta 82:337-341
92. Pinfield NJ and Davies HV (1978) Hormonal changes during after-ripening of Acer platanoides L. seeds. Z Pflanzenphysiol 90:171-181
93. Pinfield NJ, Davies HV and Stobart AK (1974) Embryo dormancy in seeds of Acer platanoides. Physiol Plant 32:268-272
94. Pinfield NJ and Stobart AK (1969) Gibberellin-stimulated nucleic acid metabolism in the cotyledons and embryonic axes of Coryllus avellana (L.) seeds. New Phytol 68:993-999
95. Pinfield NJ and Stobart AK (1972) Hormonal regulation of germination and early seedling development in Acer pseudoplatanus (L.). Planta 104:134-145
96. Pitel JA and Wang BSP (1980) A preliminary study of dormancy in Pinus albicaulis seeds. Can For Serv Bi-monthly Res 36(1):4-5
97. Rao VS, Braun JW and Khan A (1976) Promotive effects of organic solvents and kinetin on dark germination of lettuce seeds. Plant Physiol 57:446-449
98. Richardson SD (1959) Germination of Douglas-fir seed as affected by light temperature and gibberellic acid. For Sci 5:174-181
99. Ross SD, Pharis RP and Binder WD (1983) Growth regulators and conifers: their physiology and potential uses in forestry. In: Plant Growth Regulating Chemicals, Vol II Nickell LG ed. Boca Raton, Fl: CRC Press, Inc, pp 35-78

100. Rowe DCF and Gordon AG (1981) Studies on the effects of pre-chilling periods or gibberellins used to stimulate the seed germination of Nothofagus obliqua and N. procera. Seed Sci Technol 9(3):823-838
101. Salmia MA (1981) Proteinase activities in resting and germinating seeds of Scots pine, Pinus sylvestris. Physiol Plant 53:39-47
102. Salmia MA, Nyman SA, Mikola JJ (1978) Characterization of the proteinases present in germinating seeds of Scots pine, Pinus sylvestris. Physiol Plant 42(2):252-256
103. Samsonova AE (1974) Gibberellin and the growth of Picea abies seedlings Genet selektsiya i introduktsiya les porod 1:127-131
104. Shafiq Y (1980) Effect of gibberellic acid (GA_3) and pre-chilling on germination percent of Nothofagus obliqua (Mirb.) Oerst and N. procera Oerst seeds. The Indian Forester 106)(1):27-33
105. Shannon PRM, Sharrard RM and Jarvis BC (1981) Messenger RNA in embryonic axes of dormant and non-dormant seeds of hazel (Corylus avellana L.). Plant Cell Physiol 22:1293-1304
106. Shannon PRM and Jarvis BC (1982) $Poly(A)^+$RNA synthesis in cotyledons of dormant and nondormant seed of hazel (Corylus avellana L.). Plant Science Letters 26:83-91
107. Shibakusa R (1980) Effect of gibberellin A_3, IAA, ethrel and 6-benzyl-adenine on seed germination of Picea glehnii. J Japan For Soc 62(11):440-443
108. Simak M (1973) Separation of forest seed through flotation. Proc IUFRO International Symposium on Seed Processing, Vol 1, Bergen Norway, Paper No 16, 20 p
109. Simmonds JA and Dumbroff EB (1974) High energy charge as a requirement for axis elongation in response to gibberellic acid and kinetin during stratification of Acer saccharum seeds. Plant Physiol 53:91-95
110. Simola LK (1974) The ultrastucture of dry and germinating seeds of Pinus sylvestris L. Acta Botanica Fennica 103. 31 p
111. Simola, LK (1976) Changes in the subcellular organization of endosperm and radicle cells of Picea abies during germination. Z Pflanzenphysiol 78:41-51
112. Sokolov VS (1963) Effect of gibberellin solutions on the germination of seeds of tree species. Lesn Z Arhangel'sk 6(1):9-12
113. Sondheimer E, Tzou DS and Galson EC (1968) Abscisic acid levels and seed dormancy. Plant Physiol 43:1443-1447
114. Sovershaev PV and Larin VB (1971) Effect of certain physiologically active substances on nucleic acid metabolism during the germination of Norway spruce seeds. Lesno Zhurnal 14(5):5-8
115. Stearns F and Olson J (1958) Interaction of photoperiod and temperature affecting seed germination in Tsuga canadensis. Am J Bot 45:53-58
116. Steinemetz CL and Roberts BR (1984) An analysis of the gibberellic and abscisic acid content of white ash seeds. J Arboriculture 10(10):283-285
117. Tan Z, Dong Y, Fang Y and Gao R (1984) The relations of seed dormancy with abscisic acid and seed coat in Korean pine Pinus koraiensis. Sci Sin Ser B 27(2):178-186
118. Tao KL and Khan AA (1974) Penetration of dry seeds with chemicals applied in acetone. Plant Physiol 54:956-958
119. Taylor JS and Wareing PF (1979a) The effect of stratification on the endogenous levels of gibberellins and cytokinins in seeds of Douglas-fir (Pseudotsuga menziesii (Mirb. Franco)) and sugar pine (Pinus lambertiana Dougl.). Plant Cell Envir 2:165-171

120. Taylor JS and Wareing PF (1979b) The effect of light on the endogenous levels of cytokinins and gibberellins in seeds of Sitka spruce (Picea sitchensis Carriere). Plant Cell Envir 2:173-179
121. Taylorson RB and Hendricks SB (1977) Dormancy in seeds. Annu Rev Plant Physiol 28:331-354
122. Taylorson RB and Hendricks SB (1979) Overcoming dormancy in seeds with ethanol and other anesthetics. Planta 145:507-510
123. Terry ME, McGraw D and Jones RL (1982) The effect of IAA on growth and soluble cell wall polysaccharides centrifuged from pine hypocotyl sections. Plant Physiol 69:323-326
124. Thomas, TH (1977) Cyktokinins, cytokinin-active compounds and seed germination. In: The Physiology and Biochemistry of Seed Germination. Khan AA ed. Amsterdam: North-Holland Publishing Company, p 111-144
125. Tillberg E and Pinfield NJ (1981) The dynamics of indole-3-acetic acid in Acer platanoides during stratification and germination. Physiol Plant 53:34-38
126. Tompsett PB (1983) The influence of gaseous environment on the storage life of Araucaria hunsteinii seed. Ann Botany 52:229-237
127. Toole VK, Toole EH, Hendricks SB, Borthwick HA and Snow AG (1961) Responses of seeds of Pinus virginiana to light. Plant Physiol 36:285-290
128. Toole, VK, Toole EH, Borthwick HA and Snow AG (1962) Responses of seeds of Pinus taeda and P. strobus to light. Plant Physiol 37:228-233
129. Trewavas AJ (1982) Growth substance sensitivity: the limiting factor in plant development. Physiol Plant 55(1):60-72
130. Trewavas A (1986) Understanding the control of plant development and the role of growth substances. Aust J Plant Physiol 13:447-457
131. Tzou DS, Galson EC and Sondheimer E (1973) The metabolism of hormones during seed germination and release from dormancy. III. Plant Physiol 51:894-97
132. van Staden J, Forsyth C, Bergman L and von Arnold S (1986) Metabolism of benzyladenine by excised embryos of Picea abies. Physiol Plant 66:427-434
133. van Staden J, Webb DP and Wareing PF (1972) The effect of stratification on endogenous cytokinin levels in seeds of Acer saccharum. Planta 104:110-114
134. Venator CR (1972) Effect of gibberellic acid on germination of low-vigour Honduras pine seeds. For Sci 18(4):331
135. Villiers TA (1968) An autoradiographic study of the effect of the plant hormone abscisic acid on nucleic acid and protein metabolism. Planta 82:342-354
136. Villiers TA (1971) Cytological studies in dormancy. I. Embryo maturation during dormancy. New Phytol 70:751-760
137. Villiers TA and Wareing PF (1960) Interaction of growth inhibitor and a natural germination stimulator in the dormancy of Fraxinus excelsior L. Nature 185:112-114
138. Villiers TA and Wareing PF (1965) The possible role of low temperature in breaking the dormancy of seeds of Fraxinus excelsior L. J Exp Bot 16:519-531
139. Wareing PF (1982) Hormonal regulation of seed dormancy--past present and future. In: The Physiology and Biochemistry of Seed Development Dormancy and Germination. Khan AA ed. Amsterdam: North-Holland Publishing Company, pp 185-202

140. Wareing PF and Saunders PF (1971) Hormones and dormancy. Annu Rev Plant Physiol 22:261-188
141. Wcislinska B (1977) The role of gibberellin acid (GA_3) in the removal of dormancy in Fraxinus excelsior L. seeds. Biologia Plantarum 19(5):370-376
142. Webb DP (1974) Germination control of stratified sugar maple seeds. For Chron 50(1):112-113
143. Webb DP and Dumbroff EB (1969) Factors influencing the stratification process in seeds of Acer saccharum. Can J Bot 47:1555-1563
144. Webb DP, van Staden J and Wareing PF (1973a) Seed dormancy in Acer. Changes in endogenous cytokinins gibberellins and germination inhibitors during the breaking of dormancy in Acer saccharum Marsh. J Exp Bot 24:105-116
145. Webb DP, van Staden J and Wareing PF (1973b) Seed dormancy in Acer. Changes in endogenous cytokinins gibberellins and germination inhibitors during the breaking of dormancy in Acer pseudoplatanus L. J Exp Bot 24:741-750
146. Webb DP and Wareing PF (1972a) Seed dormancy in Acer pseudoplatanus: The role of the covering structures. J Exp Bot 23:813-829
147. Webb DP and Wareing PF (1972b) Seed dormancy in Acer: Endogenous germination inhibitors and dormancy in Acer pseudoplatanus L. Planta 104:115-125
148. Weiler EW (1984) Immunoassay of plant growth regulators. Annu Rev Plant Physiol 35:85-95

Plant Growth Regulation 6: 95–111

5. PLANT GROWTH REGULATORS IN CELL AND TISSUE CULTURE OF WOODY PERENNIALS

D.J. Durzan
Professor, Department of Pomology, University of California, Davis, CA 95616, USA

ABSTRACT

Plant growth regulators (PGRs) are important factors in the clonal propagation of many woody perennials. Principles involving the utility of exogenously applied PGRs are not always clearly understood. The qualitative reasoning, process control and optimization of methods using PGRs are related to a signal-theory model describing the dynamic behavior of metabolic networks and physiological states as explants proceed to discrete developmental outcomes. Algorithms that emerge from this reasoning are sought as criteria in attempts to capture genetic gains through true-to-type clonal products and in identifying when, where and how aberrations arise.

1. INTRODUCTION

This article reviews several new concepts about how plant growth regulators (PGRs) applied to explants of woody perennials facilitate clonal propagation. In all of this work, the goals are to recover trees that are true-to-type and to understand developmental processes so that some degree of mechanization and quality control can be achieved.

We now suspect that PGRs interact strongly with explant genetics, physiological states of cells, nutrition in the explant and medium, and cultural environments (11,12,44,46). We need models based on improved qualitative and quantitative reasoning to establish useful diagnostic and prognostic methods for solving problems with woody perennials.

Our understanding of the action of PGRs has recently progressed considerably with use of radioimmunoassays, mono- and polyclonal antibodies, and ELISA tests (see other chapters in this book). My review provides a brief and general framework for understanding the responsiveness of target cells and tissues in explants and in clonal products. The major challenge is to find the right cells or physiological state to exploit totipotency. The outcome often depends on controlling PGR activity through sequential treatments and cultural interactions that contribute to true-to-type development. The finding that some competent cells and tissue cultures often produce "cloudy factors," extracellular gels, and oligosaccharins opens a new chapter in the study of PGRs that is considered only in general here because of our ignorance of this area (1,13).

Over the years, we have attempted to characterize the physiological state of explants that are responsive to PGRs (e.g., 17,23,34). This approach helps to identify the useful highly totipotent explants for the development of somatic embryos (10,14,23,24). My review will deal with 1) specific problems with exogenously applied PGRs, 2) qualitative reasoning for diagnostic purposes, 3) the behavioral aspects of physiological states and metabolic phenotypes, and 4) process control and its optimization.

In any such endeavor, the arbitrary and incomplete aspects of the proposed models must be recognized. No claim is made that the concepts outlined provide a universal solution to our problems. I hope to develop the concept that an optimal PGR-directed metabolic command language or algorithm should be sought to explain the action of PGRs in cell and tissue culture systems aimed at discrete developmental outcomes.

2. EXOGENOUSLY APPLIED PGRS

At least two problems arise when dealing with exogenously applied PGRs in cell and tissue culture systems: 1) choosing the most appropriate and cost-effective PGRs in proper amounts and proportions, 2) choosing and delivering the frequency of dose. Both require some understanding of factors involved in PGR action. Each will be highlighted from the perspective of the cell and tissue culture scientist. The aim will be to establish some qualitative reasoning and concepts that derive from this reasoning.

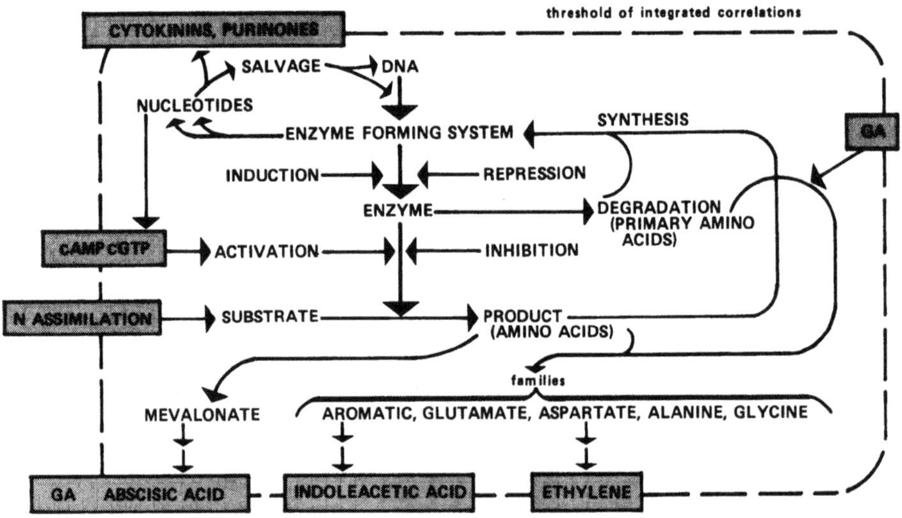

FIGURE 1. Processes (induction/repression, activation/inhibition) in enzyme-forming systems originating from information in DNA and related to nitrogen assimilation and to the biosynthesis of certain plant growth regulators (after 10). Bifurcations in metabolic networks, signals, thresholds, and stability of flux leading to overall gene expression must somehow be integrated with PGR activity. New diagnostic methods are needed to view developmental fields and metabolic phenotypes arising from this type of hierarchy in biosynthesis. Positive and negative thresholds emerging from the integration of these events, however specified, can be

viewed by computer-assisted graphics as in Fig. 3. A variation to this model is the emergent system based on enzyme activity in RNA. We do not yet know if and how PGRs affect the ribozyme-forming system (5). Regulation of events by calcium is not indicated, i.e. the relations are arbitrarily chosen and incomplete.

2.1. PGR choice and dose

For conifer micropropagation, cytokinins (e.g. kinetin, N^6-benzyladenine (BAP)),auxins (e.g., α-naphthalene acetic acid (NAA), β-indolebutyric acid (IBA), and, for somatic embryogenesis, 2,4-dichlorophenoxyacetic acid (2,4-D) and abscisic acid are currently the most commonly used PGRs. Experience has shown that these PGRs, alone or in combination, are often insufficient to program developmental outcomes for most woody perennials. Especially important are explant source, pretreatments, developmental age, defined growth media, nutritional supplements, sequential treatments, aeration, and subculture rates.

Gibberellins, abscisic acid, ethylene and more recently polyamines, are becoming increasingly important components of true-to-type process control. For example, the destruction of ethylene or other organic volatiles by $KMnO_4$ pellets in fresh humid air around the explant may avoid ethylene-induced senescence and the build-up of some polyamines. This technique can be beneficial particularly when dealing with explants of mature tissues (i.e., those from flowering woody perennials). It also underscores the general concept that growth and development represent a balance of stimulatory and inhibitory substances, many of which may yet remain unidentified.

In cell and tissue culture, one consideration is the efficiency or "coverage" that is provided by the PGR "balance" in maintaining whole plant correlations and tissue multiplication rates. Coverage is complicated by past physiological states (explant source, sensitivity, responsiveness, etc.) and by the endogenous levels of PGRs. Balance is complicated by the correlations lost upon excision of the explant, wounding reactions, levels of inhibitors, and by PGR coverage.

Usually PGRs are given to explants in specific doses and are highest at zero time when explants are placed into media. Subsequent levels of PGRs are adjusted for practical purposes by inoculation density and by the rate of subculture. "Dose frequency" is best determined on the basis of the useful half-life of the PGR or of PGR mixtures. For cell suspensions and bioreactors, greater attention has been placed on continuous infusion of PGRs. Since 1978, I have sometimes used Alzet microosmotic pumps suspended in a nipple flask to achieve a more uniform PGR dose. Uniform doses improve the synchronous development in cell suspension cultures (unpublished observations).

The common initial doses of PGRs in micropropagation of conifers are illustrated in the review by Durzan (9). The data do not reveal the half-life of the PGRs, subculture rates or dynamics of action, e.g., distribution in tissues, and peak to trough swing in concentrations. This information is not yet widely available. Many propagators believe that the initially high levels of exogenous PGRs may trigger undesirable actions (26,46). Lo- levels may simply be ineffective. As a result, most professionals, rather than having to explore continuous feeding methods, still tend to adjust by altering the subculture frequency. For this reason the appropriate dose-response curves must be developed. PGRs may be given as single doses or as continuous or continual infustion under defined conditions. The notion that emerges here is one of dependency of

PGR action on diffusion (41,42) and a critically defined schedule of delivery under specific environments for any given finite developmental expectation.

2.2. Factors affecting PGR action

Apart from specificity, the main factor affecting PGR action is dependency upon hitting a high or low peak of concentration of a PGR, if only briefly. In some cases, explants have been preconditioned in one way or another, e.g. excised from a specific position or time, dipped, or exposed briefly to high levels of PGRs to initiate morphogenesis and avoid toxic side-reactions. Another factor emerges if the response sought is dependent on the total PGR given to the explant. This factor may influence the sequence of development and final yield particularly in cell suspensions where exposure to PGRs may be more uniform (Table 1). Alternatively, the activity of one or more of the PGRs may be independent of the area underneath the concentration-time curve regardless of whether the PGR is given as a single or multiple exposure or by diffusion or infustion. More information is needed on the mechanism of action and half-life of the PGRs for each species, explant source, and culture systems that are employed.

Another unexplored factor is the rate of exposure of tissue to PGRs according to a time-varying schedule. Rates may be stage-specific as follows:

EXPLANT ⟶ COMPETENCE ⟶ DETERMINATION ⟶ DEVELOPMENT.
dedifferentiation induction differentiation

Temporal PGR requirements for a designated step are delineated by shifting explants from each type of inducing medium to a basal medium in all permutations and combinations (7). In this way, the time to complete a designated step can be delineated, mapped and related to the specific PGR or condition permitting completion of that step. This approach demonstrates the need for PGRs (promoters and inhibitors) to complete the above sequence.

The sequential simplification of culture media used by F.C. Steward (39) with embroyogenic carrot cell suspensions shows that progressively autotrophic conditions are needed to complete the recovery of plants from cells. Temporal mapping of developmental stages should include effects of media enrichment and starvation under uniform conditions.

Time-varying schedules for woody perennials should include seasonal and diurnal cycle of high-low delivery rates. These areas of ignorance and inattention will become increasingly important for manipulating explants from mature trees, and as we find somaclonal and developmental aberrations in clonal materials that should be true-to-type. An understanding of coverage of PGRs will help our assessment of clonal methods and field performance of cloned woody perennials. PGR coverage may some day have to be built into equations for capturing genetic gains in clonal forestry. But b-fore this occurs, we can anticipate that a lot of work will be needed to sort out PGR action x media x environment interactions for any given explant type and its developmental outcome in a given cultural system.

TABLE 1. Free amino acid composition of Douglas-fir cotyledon callus grown in cell suspension and in diffuse light for 30 days. The medium (1/4-strength Murashige-Skoog) was supplemented with L-arginine (400 mg/liter) and growth regulators (2.5 mg/liter NAA, and 0.1 mg/liter BAP) as indicated (±). Data represent the free amino acid composition on a mole % basis and reveal the strong interactions among arginine and plant growth regulators in terms of the distribution of amino acids and final yield. t, traces.

	- Arginine		+ Arginine	
Amino acid	- Growth regulators	+ Growth regulators	- Growth regulators	+ Growth regulators
asparagine	28.3	26.4	27.8	3.5
aspartic acid	.7	2.2	1.5	1.2
threonine	.2	.8	.1	1.1
lysine	.1	.3	t	.5
glutamine	38.7	11.7	56.1	12.6
glutamic acid	4.5	16.1	3.3	11.4
proline	6.8	1.2	4.5	1.2
γ-aminobutyrate	1.3	17.5	.5	17.2
alanine	.i	4.6	.4	13.3
leucine	.2	.4	.1	.8
isoleucine	.3	.4	.1	1.3
valine	.8	.3	.4	1.4
serine	13.2	4.5	1.1	4.7
glycine	.6	.3	.2	.2
methionine	0	0	0	0
tyrosine	.1	.1	t	.2
phenylalanine	.5	.2	.2	.4
histidine	.5	.1	.1	.2
tryptophan	.5	.1	0	.2
cysteine	0	0	0	0
arginine	.2	4.2	4.5	21.6
total μmoles/g fresh wt	9.2	8.3	64.3	12.3
cell yield (g fresh wt)	0.29	9.11	0.18	7.13

One strategy that is now commonly used is to select a well-defined medium, protocol, and PGRs. Explants from a wide range of germplasm are screened rather than attempt to empirically develop new media. This approach often reflects the ignorance of principles of nutrition, media formulation and PGR action and the rush to commercialize woody perennials through cell and tissue culture.

3. QUALITATIVE REASONING

In practice, 3 factors are needed for plant regeneration in cell and tissue culture: explant choice, medium composition and physical environment, i.e. genome, nutrition including PGRs, and environment. In reality, this list is not complete because we need models and algorithms for successful generic culture systems. New postulates are also needed to explain how plant hormones evoke a variety of signals for the completion of growth and development (6). To help us understand the coverage and action of PGRs in relation to "signal theory" (20), I have proposed an arbitrary state-network model (14) based on the hierarchical logic of biosynthetic reactions in metabolic networks of developing tissues (Figures 1 and 2). The relevant interactions among composition, structure and behavior of explants also need to be identified (Fig. 3).

Emphasis has been placed on nitrogen nutrition and metabolism (10, 14,18,40) because of the importance of proteins and nucleic acids in growth and development. The wide range of compounds in the soluble N pool are also useful in diagnostic protocols. The portrayal of N metabolism may include free amino acids, polyamines, monosubstitued guanidines, conjugates with phenols and sugars, etc. Unfortunately, resources are rarely available to explore in detail the actions of PGRs using the tools of molecular biology. Nevertheless, the approach is not unlike the clinical laboratory in hospitals that have the capability of profiling the syndromes of patients.

Frameworks for qualitative reasoning should include local area metabolic networks that are specifically responsive to PGRs in causal (deterministic and nondeterministic) ways (3). The function of the exogenously applied PGRs should relate to our understanding of the physiological state of the explant and the entire tree from which the explant was taken. Variables may be discrete or continuous and should relate directly to signals and specific actions of the PGR set (exogenous and endogenous). Variables should be defined by time-dependence or independence and ultimately associated with the sensitivity of physiological state of the explant or its products.

In our model, the ontogeny of metabolic networks is based on the control-logic of biosynthetic steps in families of amino acids in the soluble N pool and on the action of PGRs (Fig.1). Ontogeny may be constrained by variables associated with specific physiological states originally in the explant or as expressed <u>in vitro</u>. The graphic display of these metabolic activities may be akin to "adaptive landscapes" (<u>cf.</u> 45). Ontogeny becomes evident where we have a sufficiently varied genetic background to explore in our experimental material.

The frequency and probability of events, which are portrayed by computer graphics (2,28), become important from a diagnostic viewpoint. Failure or aberration of the action of a PGR is defined by a "pattern match" with the expected map behavior. Aberrations in graphic displays yield information about malfunctions and the true-to-type performance.

For true-to-type development, our reasoning requires that we isolate explants and cellular systems having equally probable fates. Hence we have sought to find simple color-coded methods to describe the developmental potential of cells as they relate to morphogenesis and use of PGRs (e.g., 34,23). We have extended our double-staining method to include the free nuclear stage of somatic polyembrogenesis (24).

For each PGR, a first step involves finding the limiting values and the relative frequencies of each cellular type and situation. Where levels of PGR are toxic, the cellular system becomes unstable so that relative frequencies of moribund cells increases dramatically. This result may coincide with the loss or aberration of signals leading to the build-up of metabolites that characterize blocks in metabolism. Ultimately, PGR performance should be related to one of several mutually exclusive system events or perturbations. In PGR coverage, we expect to find transitional physiological states, signals (e.g. thresholds), and feedback behavior leading to a finite developmental expectation (roots, shoots, embryos, etc.).

We believe that physiological state-metabolic network maps represent the summation and integration of effects of exogenous and endogenous PGRs in the developing system. Notions of this type are quite common in the PGR literature, but to our knowledge, no one has yet experimentally portrayed the "response field" (30) that PGRs determine causally.

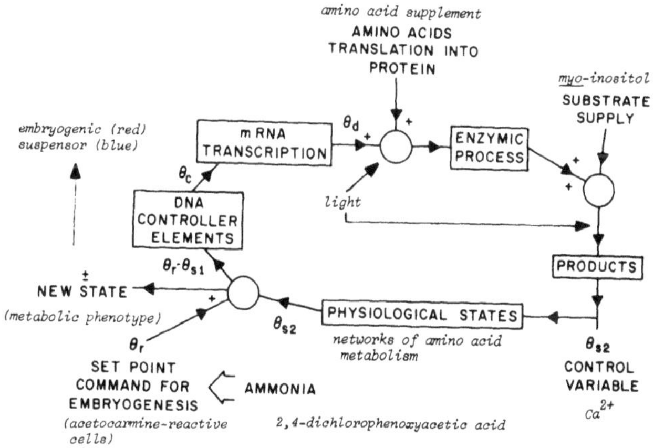

FIGURE 2. A signal-flow diagram for process control in conifer-type somatic polyembryogenesis (SPE) by 2,4-D in the presence of ammonia or reduced N. All signals are designated by θ. The set point or command signal θ_r is fed into proembryonal cells where it is compared with the current physiological state θ_{s1}. The difference between the set point and the signal θ_{s2} is called the deviation or error signal. Removal of 2,4-D comprises a forcing signal for SPE by launching proembryo development and early embryogeny. The nuclei respond and generate free-nuclear divisions which may involve amplifying, integrating or differentiating the error signal. The signal θ_c represents a RNA polymerase complex acting on DNA

which evokes mRNA production associated with the formation of embryonic proteins. Another signal based on mRNA production θ_d moves to a site of protein synthesis. This step is promoted by casein hydrolysate. A change in products produced by enzymic reactions and promoted by myo-inositol, represented as θ_{s2}, provides control over the next step in process control. This is achieved by comparison with the previous set point to define the next physiological state. Cell multiplication is maintained in darkness on media with 2,4-D and reduced N. New state refers to the developmental potential of cells that can be determined by a diagnostic double-staining color-coded procedure for the free-nuclear stage in SPE (23,34).

3.1. Oligosaccharins

While not shown in Figure 1, PGR action extends to the cell wall and its components, e.g. cellulose, hemicellulose, lignin, glycoproteins, soluble phase of walls, etc. Surveys of the literature (1,13) reveal that all components are affected by natural endogenous and exogenous plant growth regulators. More intriguing, however, has been the observation that components isolated from the cell wall have PGR properties at levels often well below the commonly used synthetic PGRs. These cell-wall components, termed oligosaccharins (1), have a very wide range of molecular structures. They represent essentially very complex mixtures of substances some of which may contain nitrogen in the form of amino acid residues or as aminosugars. Moreover, the initial fractions came from sycamore cell suspensions of D.A. Lamport which have never shown competence for development for nearly two decades (Albersheim, personal communication).

While it is too early to know much more about oligosaccharins, their unexpected importance in morphogenesis is already implicated through phenomenological observations of tissue cultures (1,13). In our efforts to evoke somatic polyembryogenesis in conifer cell suspensions, we have continually observed membranes, extracellular materials associated with early events of organogenesis (sphaeroblast, root and shoot formation), and somatic and zygotic embryogenesis (cf. 13,16,23 for illustrations). Earlier, Romberger and Tabor (36) observed complex carbohydrates deposited under excised shoot tips of *Picea abies*.

Preliminary chemical analysis of the extracellular mucilaginous material involved in somatic polyembryogenesis indicates that it has 80% alcohol-soluble and insoluble components. Cell suspensions exposed to 2,4-D and NAA produce this material and increase the osmolality of the medium sometimes by 10%. Components found in gels include glycoproteins, xylans, galactans, and several but not all free amino acids. Clearly considerable work remains to be done to fully understand the role of extracellular factors that are now postulated to mimic the action of PGRs.

FIGURE 3. Nitrogen flux in metabolic networks and daily changes in physiological states during the development of a tree fruit *in situ*: epi and mesocarp of a cling peach (Carson). Maps of this type represent true-to-type dynamics in development sought in clonal products. Implicit in these color-enhanced state-network maps (14) are underlying casual factors and correlations based on the activities of endogenous PGRs (cf. 44). Each map contains: a) logical hierarchies in the biosynthesis of amino acid families and related compounds (known and yet unidentified). b) physiological states are based on days of development and on the

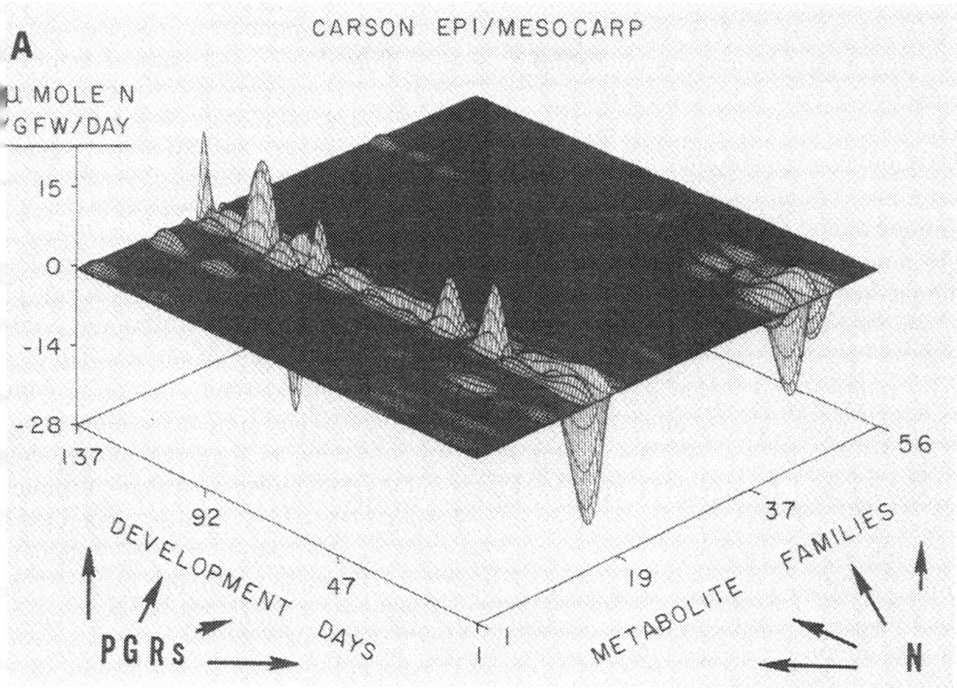
A
CARSON EPI/MESOCARP
µ MOLE N
/GFW/DAY
15
0
-14
-28
137
92
47
1
DEVELOPMENT DAYS
1
19
37
56
METABOLITE FAMILIES
PGRs
N

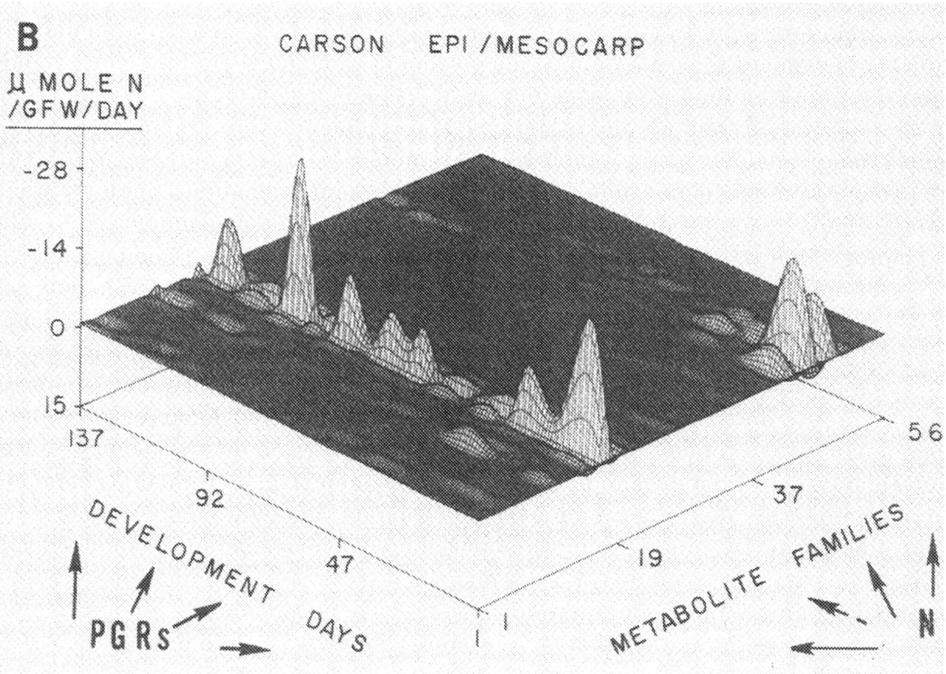
B
CARSON EPI/MESOCARP
µ MOLE N
/GFW/DAY
-28
-14
0
15
137
92
47
1
DEVELOPMENT DAYS
1
19
37
56
METABOLITE FAMILIES
PGRs
N

nation that the past history or behavior of the system affects its present and future behavior. Variables whose values represent the state are called "state variables." Both a and b have a quantitative and usually nonlinear dynamic that can be approximated mathematically. Figure 3A shows the positive departures from equilibrium of metabolite families over time. Figure 3B flips this map over to view the nagative departures over time. Threshold of various types, signals, bifurcations, stability, among other attributes, can be estimated on these maps so as to define finite and discrete developmental outcomes.

Presumably these phenotypes can be modified by the addition of exogenous PGRs. In this model the levels of coverage of endogenous PGRs was not measured. However, causal relations can be estimated by pattern matching of the metabolic phenotypes with dynamics of PGRs. Patterns leading to the developmental outcome in situ or in vitro can be matched and classified as being true-to-type or aberrant.

Recently, some attention has been given to the role of crystalline genes especially in notions dealing with the origin of life (4). The plant cell wall also has crystalline properties and may have some of the primitive replicating properties attributable to the remarkable crystalline genes of geological origin. The speculative hypothesis emerges that should the cell wall have template-replicating properties as in the formation of micelles, then the oligosaccharins, or some fragmentary component containing oligosaccharins, are tantamount to messengers in cell-wall-regulated developmental events if they create bifurcations in system behavior. The products of oligosaccharin activity are not yet known like many of the gene products in current work in molecular genetics. Because of the large number of oligosaccharin components (over 100), we can anticipate that products and activities of oligosaccharins are combinatorially enormous. The significance of oligosaccharin and PGR interactions may reside in controlling and coordinating developmental events originating in the nucleus and cell wall.

3.2. The Soluble Phase of Cells

Few of the internal subcellular compartments of cells are as complex as the soluble phase with its ionic and potentially covalent interactions. We have never had good ways to understand the soluble phase according to its compartments, dynamics, and behavioral patterns in response to PGRs. Ideally the soluble phase should be conceptualized in relation to subcellular organization, PGR receptors, transporters and loci of activity in a coordinated system of bioenergetics and membrane transformations (e.g. 30).

Metabolic models do not yet explain completely or specify exactly the mechanisms of action of PGRs like (±)-abscisic acid (27). Rather they provide the nondeterministic causal framework for the physiological states that emerge (3). Models of the soluble phase also do not yet account for the very short-term process and their specific effects such as gravi- or geotropism. More biophysical understanding and integration with the biochemistry of these systems is required.

We can anticipate many new correlations to emerge from PGR activity. Correlations will inevitably involve the large number of still unidentified naturally occurring compounds in the physiological fluids of developing plants (e.g., Fig. 4).

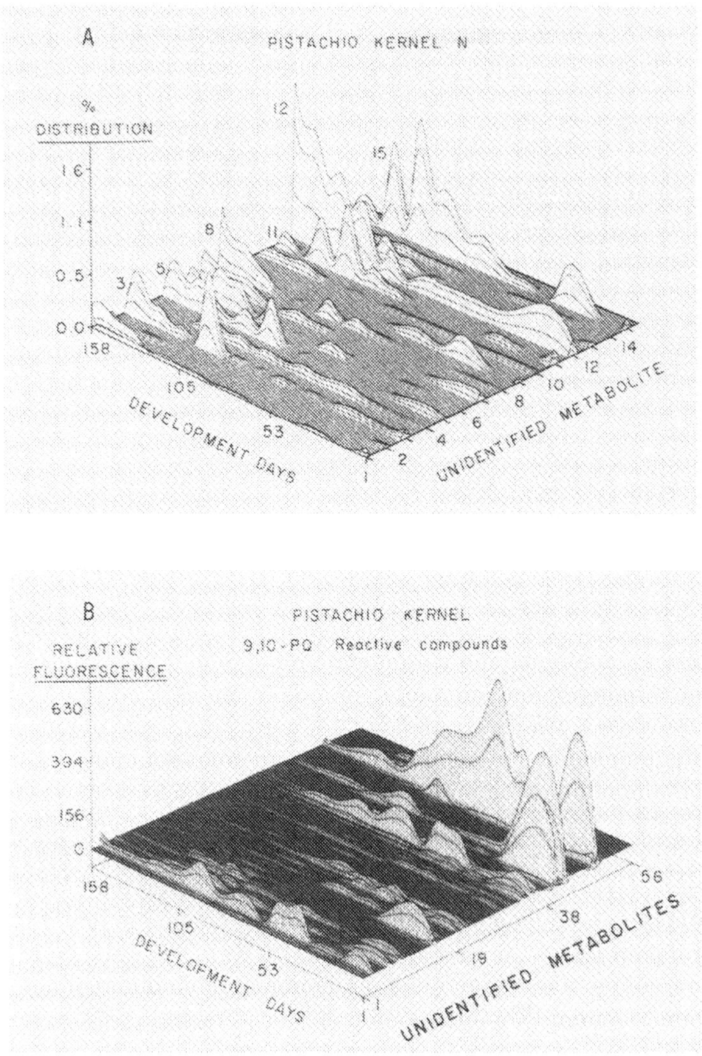

FIGURE 4. The dynamics of behavior of yet unidentified nitrogenous compounds in the developing kernel of Pistacia vera. We do not yet know if any of these compounds have PGR activity or if they regulate the global relatedness required for PGR process control in morphogenesis.
A. Percent distribution of 15 ninhydrin-reactive unidentified compounds.
B. Relative fluorescence of 56 autofluorescing and 9,10-phenanthrequinone-reactive compounds separated on an automated amino acid analyzer for physiological fluids (43).

3.3. Genetically modified cells

The Ti-plasmid of *A. tumefaciens* is now an established vector for the genetic engineering of plants. *A. tumefaciens* has a broad host range and infects a wide range of gymnosperms and dicotyledonous angiosperms. Normally, the transfer of a defined segment of the Ti-plasmid (T-DNA) into the genome of a plant results in the formation of a crown-gall tumor. However, morphologically normal plants can be produced if the tumor-inducing genes are removed (disarmed Ti-plasmid) from the transferred DNA (T-DNA). Root and shoot development in some explants can also be specified by the integration of Ti-plasmids (29). Morphogenesis is related to plasmid-directed synthesis of auxin and cytokinins. However, Owens (31) has found that with gene transfers to induce morphogenesis in alfalfa, endogenous auxin levels do not explain why morphogenesis occurs in cells. Regulation of morphogenesis involves other levels of complexity and involves gene promoter activity (foreign gene), enzyme biosynthesis and network environments and physiological states (*cf.* Figs. 1,2,3).

Agrobacterium-induced transformations have now been achieved in forest trees (e.g., 8,32,37). In micropropagated shoots from a 60-year-old Douglas-fir and in seedlings, we have demonstrated transformation by two strains of *Agrobacterium* and expression of a foreign gene (8). Synthesis of octopine was demonstrated together with the expression of the foreign gene by synthesis of the enzyme (kanamycin phosphotransferase) that endows tolerance to kanamycin added to the medium. The integration of genetic engineering methods with micropropagation and with traditional breeding represents a major step forward in capturing genetic gains and introducing new genetic information. This work enables domestication and the insertion of potentially useful foreign genes in one of North America's most valuable forest trees.

In future, the cell and tissue culture scientist will benefit enormously from the use of foreign genes as probes for the study of PGRs and development.

4. PROCESS CONTROL AND OPTIMIZATION

Development may be represented arbitrarily by a temporal series of discrete, cooperating, physiological states based on PGR-directed metabolic networks. For amino acid metabolism, this perspective governing fruit epi/mesocarp development is shown in Fig. 3. This approach alone does not do justice to the complexity of morphogenesis nor to the concept of totipotency (12,18,39). It may be insufficient to distinguish clearly the significant and critical patterns of growth and development or of specific physiological processes (e.g., 21,22,33,46). We are not sure how PGRs contribute to process control as displayed by the flux of amino acid N. It is again difficult to see how morphogenesis achieves very large scale integration and global relatedness in development. For these reasons, notions of optimal control (linear and nonlinear) are very preliminary.

4.1. Optimization of systems and PGRs

With the heirarchical relationships among Figures 1, 2 and 3, one is always looking for better methods of the real world. The real world often involves nonlinear variables. Diagnostic reasoning in nonlinear behavior requires some knowledge of computational and combinatorial mathematics. Without going into the mathematical background, my laboratory has approached metabolic-PGR system optimization by characterizing and localizing maxima and minima of the non-linear functions.

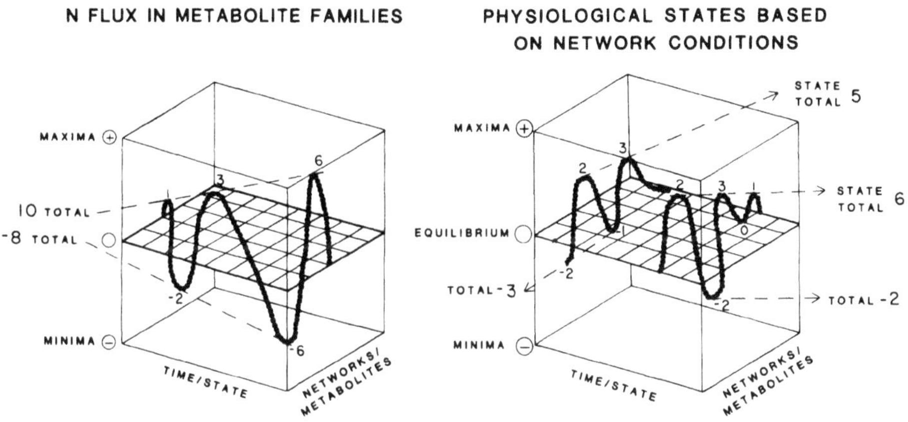

FIGURE 5. Costs of biosynthesis (A) and establishing physiological states (B) using maximal and minimal values (thresholds, signals) of the metabolic flux. Numerical values for maxima and minima are totalled over each function to simplify computations and to formulate optimal "process control" algorithms.

In Figure 5 optimization is based on threshold maxima and minima costs. One strategy the explant may employ in reaching clonal fidelity is to minimize threshold costs as for as possible. Threshold values are added to obtain global values. If there is a unique effect arising from PGR action, which leads to the lowest value, we call this value a global minimum. For the developing cling peach, the local minimum in Figure 3B is -28 μmoles N per gram of epi/mesocarp per day.

The maxima and minima of functions describing flux in the state-network map are postulated to represent constraints imposed by sets of PGRs. Hence, Figure 3 may represent a PGR-constrained nonlinear optimization problem based on an inherent metabolic command language. Some of the maxima and minima may not have PGR-activity constraints. This possibility can be tested by perturbing the system with PGRs and reconstructing diagnostic maps. With specially designed computer software, maps can be subtracted from one another to identify differences in characteristic responses. Isotopically labelled substrates can facilitate the monitoring of the metabolism by showing the dispersion of PGR products. Methods of cell and tissue culture provide tools to simplify the situation provided that we understand each step of the process.

Ultimately, constraints imposed by PGRs must be specified as a set of functions collected together to give the specified end result or phenomenon. The end result should have some repeatable values that are equal or unequal in satisfying the phenomenon. Any effect satisfying all constraints is said to be fully feasible. The set of all feasible points defines the feasible region on the state-network map. All other points are said to be infeasible. While much more could be said about these notions, the reader will recognize that our approach quickly exceeds all current experimental data.

Once we have sufficient detail and the proper correlations, the time axis of state-network maps may be more properly represented by "processes" or by algorithms for morphogenesis, e.g., gradients, positional information, and bifurcations related to the action of PGRs. More accurate geometric descriptions of global morphogenesis become spatially complicated (22). An understanding of process control for global morphogenesis and its associated algorithms will require so much more data that computerized displays will become essential (2).

Even after reliable models are available, a complete description of global relatedness among phenotypes, processes, and components may be very difficult to obtain easily and it may not always be necessary (35). Through a sampling of "critical features," the truly effective metabolic phenotypes may be discerned. The "effective sampling index" should be identified by the number of features or attributes sampled over all known variables and components. Also, some measure will be needed of the effective map configurations as opposed to the total possible configurations. Considerable knowledge, simulation, and familiarity with the experimental material will be required to achieve an understanding of process control and global relatedness.

4.2. Elite traits in clonal systems: Feed-forward and feed-back phasic behavior

A problem in capturing genetic gains with current micropropagation based on explants from embryonic or juvenile plants is the requirement for correlations with performances of the mature tree. Methods are needed to speedup the evaluation of progeny cloned from proven mature trees. This problem is accentuated in pomology where the crop derives from reproductive maturity.

Through cell and tissue culture, the totipotency of mature trees can be directed by a phase shift back to the juvenile and embryonic states or by a shift forward to maturity and senescence. For example, cell suspensions of a petiole callus of a 28-year-old cherry rootstock (*Prunus cerasus* cv. Vladimir) can be induced with PGRs to produce roots (organogenesis) and low frequency somatic embryogenesis under defined conditions (25). By contrast, the same cell suspensions can be directed to age and behave like the ripening fruit of the tree, i.e. cells behave not as meristemic tissue but senesce with the production of bright red color and fragrances characteristic of mesocarp cells of the ripening fruit (15).

The above indicates that at a specific point in the life cycle of the tree and with the physiological states of some cells we have the opportunity to move forward or back in accelerating phasic behavior. This bidirectional expression of totipotency may some day be useful in evaluating the potential for clonal propagation (rooting, micropropagation, somatic embryogenesis) or the potential for a quality crop (e.g. flavor, color, gene x environment interactions) using the same cell suspension culture.

In feedback and feedforward phase changes, PGRs are a dominant parameter, but as we have seen, not the only factor. Through greater understanding of the biology, physiology and biochemistry of woody perennials, biotechnology should enable us to evaluate the potential of germplasm far more thoroughly than ever before. However, the cost-effectiveness of this approach needs to be demonstrated. Given the current state of research with woody perennials we are probably two decades away from putting such models into practice.

5. CONCLUSION

New models are needed for a more comprehensive understanding of how PGRs relate to cell and tissue culture technologies. For the plant propagator, models are essentially protocols that represent an array of end uses. We have yet to discover the appropriate algorithms and command language that account for the action and specificity of PGRs in the short and long term. Progress will emerge with improved control over genetics of cell and tissue culture systems and with the development of more highly refined and specific continuous monitoring methods.

REFERENCES

1. Albersheim P, Darvill AG: Oligosaccharins. Amer Sci 253:58-64, 1985.
2. Brown CM: Computer vision and natural constraints. Science 224: 1299-1304, 1984.
3. Bunge M: Causality, chance and laws. Am Sci 49:432-448, 1960.
4. Cairns-Smith A, Braterman P: Search for crystal genes. Proc Intl Conf Origin of Life, Berkeley, CA, July 21-25, 1986.
5. Cech TR, Bass BL: Biological catalysis by RNA. Ann Rev Biochem 55:599-627, 1986.
6. Canny MJ: Ashby's law and the pursuit of plant hormones: a critique of accepted dogmas, using the concept of variety. Aust J Plant Physiol 12:1-7, 1985.
7. Christianson ML: Organogenesis *in vitro* as a developmental process. Proc XXII Intl Hort Congress, Aug 10-18, 1986, Univ. of California, Davis, CA (in press), 1987.
8. Dandekar AM, Gupta PK, Durzan DJ, Knauf V: Genetic transformation and foreign gene expression in micropropagated elite Douglas-fir (*Pseudotsuga meniesii*) (unpublished observations).
9. Durzan DJ: Cell and Tissue Culture in Forest Industry. *In* Tissue Culture in Forestry. Bonga JM, Durzan DJ, eds Martinus Nijhoff/Dr W Junk, The Hague, pp 36-71, 1982.
10. Durzan DJ: Nitrogen metabolism and vegetative propagation of forest trees. *In* Tissue Culture in Forestry. Bonga JM, Durzan DJ, eds Martinus Nijhoff/Dr W Junk, The Hague, pp 256-324, 1982.
11. Durzan DJ: Special problems: Adult vs. juvenile explants, Chap. 17. *In* Handbook of plant cell culture, crop species. Sharp WR, Evans DA, Ammirato PV, Yamada Y, (eds): MacMillan Publishing Company, New York, 2:472-503, 1984.
12. Durzan DJ: Potential for genetic manipulation of forest trees: Totipotency, somaclonal aberration, and trueness to type. *In* Proceedings International Symposium on Recent Advances in Forest Biotechnology. July 10-13, 1984, Michigan Biotechnology Institute, Michigan State University, Traverse City, Michigan, pp 104-125, 1984.
13. Durzan DJ: Nitrogenous constituents extracellular to the lignocellulosic cell wall. *In* National Products of Woody Plants. Rowe J, ed, Springer Verlag, NY (in press), 1987.

14. Durzan DJ: Physiological states and metabolic phenotypes in embryonic development. In Cell and Tissue Culture in Forestry, Bonga JM and Durzan DJ, eds, Martinus Nijhoff, Dordrecht, Vol 2, 405-439, 1986.
15. Durzan DJ: Plant growth regulator-directed phase specificity in cell and tissue culture for tree improvement. Proc. Soc. Am. Foresters, National Convention, Birmingham, Alabama, Oct. 8 (in press) 1987.
16. Durzan DJ: Ammonia: its analogues, metabolic products and site of action in somatic embryogenesis. In Cell and Tissue Culture in Forestry. Bonga JM, Durzan DJ, eds, Martinus Nijhoff, Dordrecht, Vol 2, 92-136, 1986.
17. Durzan DJ, Chalupa V, Mia AJ: Growth and metabolism of cells and tissue of jack pine (Pinus banksiana Lamb.) 1. The establishment and some characteristics of a proliferated callus from jack pine seedlings. Can J Bot 54:437-445,1976.
18. Durzan DJ, Steward FC: Nitrogen metabolism. In Steward FC, ed, Plant Physiology, An Advanced Treatise, Academic Press, New York, VIIII:55-265, 1983.
19. Durzan DJ, Uriu K: Metabolic networks in developing pistachio embryos (Pistacia vera cv Kerman). In Progress Developmental Biology, Part A. Slavkin HC ed, Proc Int Soc Dev Biol, 10th Int Cong ISDB, Aug. 4-9, 1985, Los Angeles, California, AR Liss Inc, NY 199-202, 1986.
20. Gardner WA: Introduction to random processes with applications to signals and systems. MacMillan, New York, 1986.
21. Garner WR: Good patterns have few alternatives. Am Sci 58: 34-42, 1970.
22. Green PB: Organogenesis. A biophysical view. Annu Rev Plant Physiol 31:51-82, 1980.
23. Gupta PK, Durzan DJ: Somatic polyembryogenesis from callus of mature sugar pine embryos. Bio/Technology 4:643-645, 1986.
24. Gupta PK, Durzan DJ: Biotechnology of somatic embryogenesis in loblolly pine. Bio/Technology 5: (in press), 1987.
25. Hansen KC, Durzan DJ: Somatic embryogenesis and morphogenesis in Prunus cerasus. Proc XXII Intl Hort Cong, Univ California, Davis, Abstr No 1130, 1986.
26. Hartmann HT, Kester DE: Plant propagation: principles and practices, 4th edition, Prentice-Hall, Inc, New Jersey, 1983.
27. Hornberg C, Weiler EW: High-affinity binding sites for abscisic acid on the plasmalemma of Vicia faba guard cells. Nature 310: 321-324, 1984.
28. Jurs PC: Pattern recognition used to investigate multivariate data in analytical chemistry. Science, 232:1219-1224, 1986.
29. Morris RO: Genes specifying auxin and cytokinin biosynthesis in phytopathogens. Ann Rev Plant Physiol 37:509-538, 1986.
30. Mühlethaler K: The ultrastructure of cells. In Historical and current aspects of plant physiology, Davies PJ (ed.), NY State Coll Agric., Cornell Univ, Ithaca, NY, 226-242, 1975.
31. Owens L: Gene transfer to induce morphogenesis. Proc XXII Intl Hort Cong, Aug 10-18, 1986, University of California, Davis, CA (in press), 1987.
32. Parsons TJ, Sinkar VP, Stettler RF, Nester EW, Gordon MP: Transformation of poplar by Agrobacterium tumefaciens. Bio/Technology 4:533-536, 1986.

33. Platt JR: Properties of large molecules that go beyond the properties of their chemical sub-groups. J Theor Biol 1:342-358, 1961.
34. Powledge T: Biotechnology touches the forest. Bio/Technology 2:762-772, 1984.
35. Quastler H: Information theory of biological integration. Am Nat 93:245-254, 1959.
36. Romberger JA, Tabor CA: The Picea abies shoot apical meristem in culture. II. Deposition of polysaccharides and lignin-like substances beneath cultures. Am J Bot 62:660-717, 1975.
37. Sederoff R, Stomp AM, Chilton WS, Moore LW: Gene transfer into loblolly pine by Agrobacterium tumefaciens. Bio/Technology 4: 647-649, 1986.
38. Steel RGD, Torrie JH: Principles and procedures of statistics. McGraw-Hill, New York, 1960.
39. Steward FC: Growth and Organization in Plants. Addison-Wesley Publ Co, Reading, Mass, 1968.
40. Steward FC, Durzan DJ: Metabolism of nitrogenous compounds. In Plant Physiology, an advanced treatise. Steward FC, ed, Academic Press, New York, 4:379-686, 1965.
41. Theologis A: Rapid gene regulation by auxin. Annu Rev Plant Physiol 37:407-438, 1986.
42. Veen AH, Lindenmayer A: Diffusion mechanism for phyllotaxis. Theoretical physico-chemical and computer study. Plant Physiol 60:127-139, 1977.
43. Ventimiglia F, Durzan DJ: The determination of monosubstituted guanidines using a dedicated amino acid analyzer. Liquid Chromatography/Gas Chromatography 4:1121-1124, 1987.
44. Westwood MN: Temperature-zone pomology. WH Freeman and Co, San Francisco, 1978.
45. Wright S: Evolution and the genetics of populations. Univ Chicago Press, Vols 1 to 4, 1968.
46. Zaerr J, Mapes MO: Action of growth regulators. In Tissue Culture in Forestry. Bonga JM, Durzan DJ, eds, Martinus Nijhoff/Dr W Junk, The Hague, pp 231-255, 1982.

Plant Growth Regulation 6: 113-135

6. PLANT GROWTH REGULATOR EFFECTS IN THE *IN VITRO* PROPAGATION OF THREE HARDWOOD TREE GENERA: *CASTANEA*, *JUGLANS*, AND *QUERCUS*

O.J.SCHWARZ, Associate Professor, Botany Department, University of Tennessee, Knoxville 37996-1100

ABSTRACT

Castanea, *Juglans*, and *Quercus* are genera containing hardwood forest species that are difficult to propagate asexually by conventional means. Their *in vitro* clonal propagation would assist forest tree improvement and silvacultural research. The research efforts in clonal propagation of these three genera are reviewed with respect to plant growth regulator effects as they are related to primary explant source, multiple shoot formation, somatic embryogenesis, rooting, and transfer to soil.

1. INTRODUCTION

This paper reviews the use of plant growth regulators in the *in vitro* culture of hardwood species found in three genera known to be difficult to vegetatively propagate by conventional methods: *Castanea*, *Juglans*, and *Quercus* (75, 83). Various hardwood explant types derived from juvenile to adult trees are being cultured *in vitro* in order to produce clonal progeny. Over the last few years, there have been numerous books and reviews focusing on various aspects of the application and efficacy of clonal propagation of hardwoods by *in vitro* methodologies (3, 9, 19, 20, 64, 65, 70, 81). The rational for much of the *in vitro* hardwood research efforts now underway seems to be at least two-fold in nature: first, to capture the maximum genetic gain available through the cloning of superior genotypes for use in forest tree improvement programs (3, 40) and, second, to develop culture systems capable of mass propagation of species that are difficult to asexually propagate using conventional methods (3, 34). Through the use of selected cloned plantlets, genetic gains have been projected of at least 10% over those attainable by utilization of seedlings obtained from selected half-sib families (21, 36).

Brown and Sommer (10) list three major types of culture systems that have led to the regeneration of whole plants: (1) organ cultures, (2) callus cultures, and (3) cell suspension cultures. Each of these cultural approaches is well represented in the *in vitro* hardwood literature (2, 10, 65, 70). Sommer and Wetzstein (65) list some 42 genera in which at least 104 species and varieties of woody dicotyledonous trees have been regenerated via *in vitro* culture. Figure 1 illustrates several of the major routes to multiplication and regeneration of hardwoods currently being investigated. The morphogenic "circuitry" illustrated in the figure is by no means exhaustive, as a perusal of the past and current literature will quickly point out. In general, however, the propagative processes illustrated in the figure can follow two major pathways: either direct or indirect organogenesis (28).

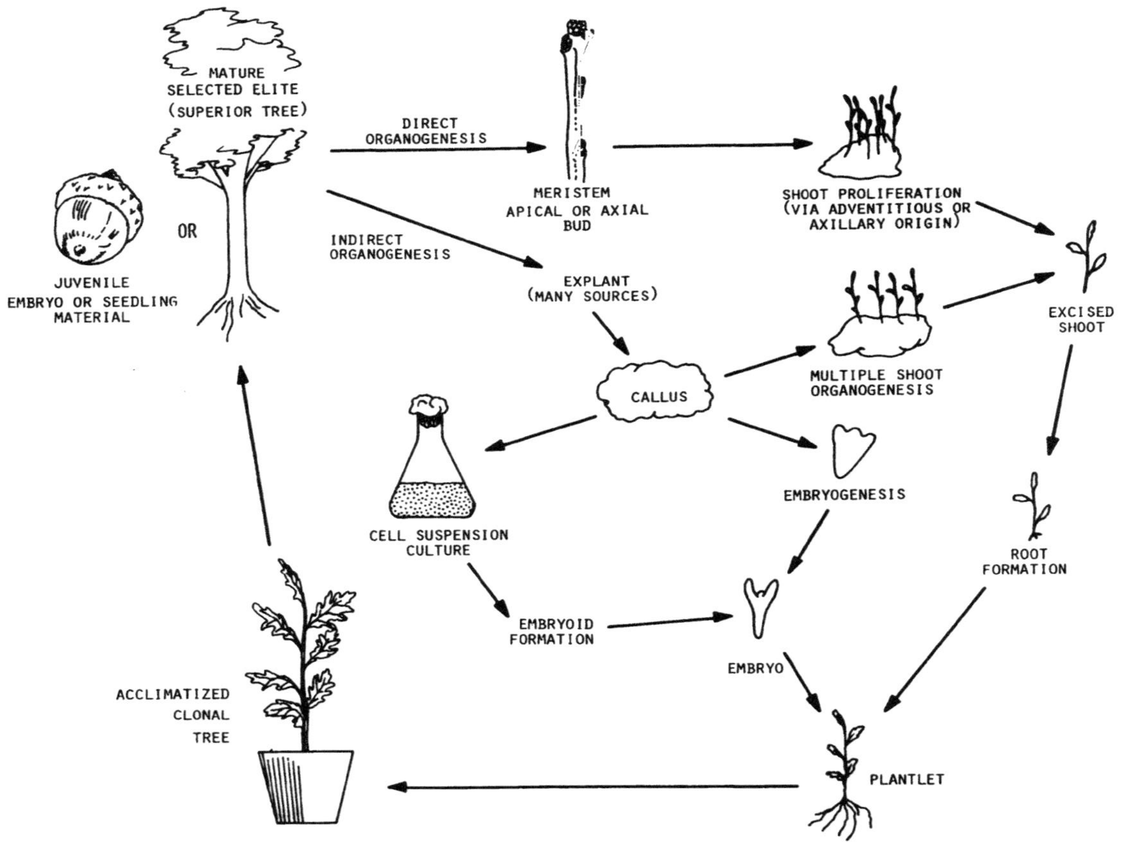

FIGURE 1. Schematic representation of two major routes of clonal propagation of hardwood trees starting with juvenile or mature explant material.

Indirect organogenesis involves the production of callus from the primary explant followed by the appearance of meristemoids (73) [i.e., "...a meristem-like aggregation of smaller, isodiametric, thinner-walled, micro-vacuolated cells which have densely staining nuclei and cytoplasm." (71)]. These meristemoids are initially not determined (28) and are capable of developing into either shoot or root primordia (58). A variation of this developmental route involves the production of complete embryos directly from the callus (i.e., somatic embryogenesis) (4). A second subroute to plantlet formation via somatic embryogenesis involves the transfer of the callus to liquid media to produce cell suspension cultures. These cell suspensions contain single cells and small aggregates of cells that can be manipulated to produce a series of embryo-like structures that, under the appropriate conditions, develop into complete embryos and eventually plantlets (72). A point worth mentioning here is that the incorporation of a callus and/or liquid suspension stage in the process of clonal propagation introduces an increased possibility of variation in the chromosomal constitution of the cells and, hence, the resultant plantlets (5, 16, 34, 37, 61).

Direct organogenesis (Fig. 1) is characterized by the absence of an intervening callus stage between the primary explant (in this case apical or axillary buds) and either direct organ production or meristemoid formation giving rise to organ formation (28). This route to clonal propagation has several advantages: (1) normal organization of the cultured explant is more or less maintained, avoiding passage through a dedifferentiated callus state and thus reducing the risk of producing aberrant progeny (30); (2) appropriately isolated "true" meristem explants can provide a source of virus-free plants (67, 69); and (3) an opportunity is provided for the preservation of germplasm (69). Clonal propagation via shoot-tip or bud culture (i.e., apical or axial) is widely used in the commercial production of woody and herbaceous plants (45, 47, 69, 83). When appropriately cultured, a mass of microshoots can be produced. The microshoots may arise spontaneously (1, 17) or as a result of exogenous application of plant growth regulators. They may be of adventitious origin (30) or they may appear as axillary shoots induced by the depression of apical dominance by exogenously applied cytokinin (1, 2, 69).

The successful utilization of all of these approaches for the clonal propagation of a hardwood species requires careful control of a complex set of chemical and physical variables that comprise the _in vitro_ explant environment (10, 64, 70, 81). Perhaps of equal importance is the developmental state of the explant at the time of culture (41). The fact that exogenously applied growth regulators play a critical role in facilitating the morphogenetic changes required to produce a cloned hardwood propagule is well documented in the literature (2, 25, 64, 65, 82, 83). The remainder of this paper will review the published work available to the author concerning the role that growth regulators play in the _in vitro_ propagation of hardwood species of three genera of economic and ecological importance to forestry. The three genera are of particular interest for several reasons: (1) as mentioned earlier, they all contain species that are relatively difficult to propagate vegetatively by conventional means; (2) they represent species in which forest geneticists have identified outstanding phenotypes (i.e., plus trees), selected in natural stands and plantations, whose cloning would bring immediate genetic gain (40); and (3) they represent a body of work that is relatively recent to the literature and has not been extensively reviewed. Information taken from the reviewed literature pertinent to plant growth regulator effects with respect to source of primary explant, results of treatment, and successful transfer to soil is presented in the following three sections.

2. CASTANEA

Results concerning the *in vitro* culture of *Castanea* are summarized in Table 1. Dormant winter buds of *C. dentata* of unstated age were placed into culture and treated with various auxins and cytokinin (29). The resultant callus appeared only with auxin treatment. Its growth was enhanced and its morphology was changed in the presence of kinetin. The addition of ethrel was found to promote cultural uniformity. McPheeters *et al.* (42) reported multiple shoot formation when embryos were treated with N^6-benzyladenine (BA) (2 mg/l) and 1-naphthaleneacetic acid (NAA) (0.1 mg/l). Cotyledonary explants formed embryoid-like protuberances when treated with the same hormonal mix. Limited axillary bud proliferation occurred when stem tip explants were cultured in the same hormonal medium. Stem sections cultured on basal medium without hormones produced callus.

Vieitez and Vieitez (76, 77) produced axillary shoot multiplication from both embryonic axes and lateral buds of 3- to 4-month-old *C. sativa* seedlings by treatment with BA. Both explant types required BA at 1 mg/l to establish the cultures, followed by a tenfold reduced dose (0.1 mg/l) to facilitate shoot multiplication. Roots were initiated on the shoots produced from the embryonic material by an 8-day treatment with indole-3-butyric acid (IBA) (1 mg/l), followed by transfer to a hormone-free basal medium. Fifty percent of the shoots rooted after 8 weeks. The transfer to IBA-free medium was said to avoid the inhibition of subsequent root growth by IBA and also the production of an inhibitory black exudate that accumulated in the solidified medium in IBA's presence.

Plantlets of *C. dentata* were regenerated (35) from mature stratified embryos by a two-step shoot multiplication procedure similar to that reported for *C. sativa* (77). Seeds were stratified in moist sand in plastic bags at 34C for 60 days. If stratified embryos were used, the percentage of cultures developing axillary buds doubled from 23 to 41%. Axillary bud development was initiated by culture in BA (1 mg/l) for 6 weeks. The explants were then transferred to a modified basal medium containing reduced nitrates and BA at 0.1 mg/l. After 4 to 6 weeks, microshoots were placed on media containing IBA at 1 or 5 mg/l. All shoots on IBA at 5 mg/l callused but produced no roots. Shoots treated with 1 mg/l IBA produced less callus, and 2 of 31 individuals produced roots that developed not from the callus mass at the base of the shoot but from the stem above the callus. A single plant was transferred to greenhouse conditions and had survived for 4 weeks.

Plantlet regeneration of *C. sativa* was accomplished using apical bud cultures derived from seedling material (55). Explants were initially cultured on a basal medium containing BA at 5 mg/l. Cultures were maintained on an 18-h photoperiod at 25C. The explants formed multiple shoots, from which meristem tips (0.3 x 0.2 mm) were excised and cultured on medium containing BA at 1 mg/l plus IBA at 0.1 mg/l for 2 weeks. The cultures were then transferred to various hormonal media designed to maximize shoot proliferation. The optimal treatment was BA at 1 mg/l plus IBA at 0.01 mg/l. Three to five shoots developed directly from each meristem tip. After 20 days, shoots longer than 2 cm were transferred to rooting medium. Rooting occurred within 20 to 25 days in half-strength

TABLE 1. Castanea

Species	Maturity	Primary Explant	Result	Growth Regulator and Effective Concentration	Successful Transfer to Soil	Reference
sativa	J	embryonic axes	axillary shoots	BA (1 to 2 mg/1)	--	Vieitez and Vieitez, 1980a
			shoot multiplication	BA (0.1 mg/1)		
			root initiation	IBA (1 mg/1)		
sativa	J	lateral buds of 3- to 4-month-old seedlings	axillary shoots	BA (1 mg/1)	--	Vieitez and Vieitez, 1980b
			shoot multiplication	BA (0.1 mg/1)		
sativa	J	seedling shoot tips	multiple shoots	BA (5 mg/1)		Rodriguez, 1982
			shoot multiplication	BA (1 mg/1) + IBA (0.01 mg/1)		
			root initiation	IBA (1 mg/1)	yes	
sativa	J	axillary buds of 3-month-old seedlings	multiple shoots	BA (1 mg/1)		Vieitez and Vieitez, 1982/83
			shoot multiplication	BA (0.1 to 0.5 mg/1)		
		embryonic axes	multiple shoots	BA (1 mg/1)		
			shoot multiplication	BA (0.1 to 0.5 mg/1)		
			root initiation	IBA (3 mg/1) IBA (1 mg/1) (2-min. dip)	yes	
sativa	J	axillary buds of 3-month-old seedlings	multiple shoots	BA (1 mg/1)	--	Chevre et al., 1983
			shoot multiplication	BA (1 mg/1) alone or in combination with IBA (0.1 mg/1)		
			root initiation	IBA (1 g/1) (20-sec. dip)		
	M	axillary buds of 5-year-old trees	multiple shoots	BA (1 mg/1)		
sativa x crenata	M	shoot tips (stump sprouts, 10- to 18-year-old trees)	shoot and axillary bud elongation	BA (1 mg/1)		Vieitez et al., 1983
			shoot multiplication	BA (0.2 mg/1)		
			root initiation	IBA (1 g/1) (2-min. dip)	inferred	

TABLE 1. Castanea (Continued)

Species	Maturity	Primary Explant	Result	Growth Regulator and Effective Concentration	Successful Transfer to Soil	Reference
sativa	J	epicotyl sections	adventitious buds	BA (2 mg/l) + IBA or NAA (0.01 mg/l)		San Jose et al., 1984
			shoot multiplication	BA (0.1 mg/l)		
			root initiation	IBA (2 mg/l) or IBA (0.5 to 1.0 g/l) (15- or 2-min. dip, respectively)	--	
		hypocotyl sections (3- to 5-week-old seedlings in vitro)	adventitious roots	NAA or IBA (0.1 mg/l)		
sativa	J	epicotyl sections	adventitious buds	BA (1 to 2 mg/l) + IBA or NAA (0.01 mg/l)	--	San Jose and Vieitez, 1984
		hypocotyl sections	adventitious roots	NAA or IBA (0.1 mg/l)		
sativa x crenata	J	cotyledons	somatic embryogenesis: globular, heart-shaped, and torpedo-shaped embryoids, no viable plants	2,4-D (0.1 to 5 mg/l) alone or in combination with various concentrations of BA, kinetin, or zeatin (0.1 to 1.0 mg/l)	--	Gonzalez et al., 1985
sativa	J	apical and axial buds of 4-week-old, in vitro grown seedlings	cuttings were grown and rooted on basal medium without hormones	--	--	Strullu et al., 1986
dentata	--	apical buds (dormant, winter)	callus	auxin alone or in combination with kinetin (0.1 to 1 ppm) ethrel added (produced more uniform callus)	--	Hu and Scrivani, 1977
dentata	J	embryo	multiple shoots	BA (2 mg/l) + NAA (0.1 mg/l)	--	McPheeters et al., 1980
		cotyledon	embryoid-like protuberances	BA (2 mg/l) + NAA (0.1 mg/l)		
		testa	--			
	M	shoot tip	axillary bud proliferation	--		
		stem	callus			
		bur	--			
		leaf	--			

TABLE 1. Castanea (Continued)

Species	Maturity	Primary Explant	Result	Growth Regulator and Effective Concentration	Successful Transfer to Soil	Reference
dentata	J	embryos	axillary shoots	BA (1 mg/l)	--	Keys and Cech, 1981
			shoot multiplication	BA (0.1 mg/l)		
			root initiation	IBA (1 mg/l)		
dentata	M	apical and lateral buds (from winter-dormant, 25-year-old trees)	multiple shoots	BA (0.5 mg/l)		Read *et al.*, 1985
			root initiation	IBA (3 g/l) (1-sec. dip)	yes	

basal medium containing IBA (1 mg/l). Higher levels of IBA (2, 5, 10 mg/l) also produced rooting; however, the resultant plants exhibited poor shoot growth, necrosis, and malformed growing tips. Plants were established in soil and grew normally.

Vieitez and Vieitez (78) reported similar results using embryonic axes and 1-cm nodal sections from 3-month-old seedlings of C. sativa to regenerate plantlets. These explants were first cultured on basal medium containing BA at 1 mg/l (76, 77). The resulting shoots were used for shoot multiplication and rooting experiments. BA was tested at five concentrations (0.05, 0.1, 0.5, 1, 2 mg/l) and the mean shoot multiplication was determined. BA concentrations of 0.1 to 0.5 mg/l gave maximum proliferation as well as the healthy growth characteristics found necessary for subsequent rooting. Both NAA and IBA were tested in the rooting experiments. NAA was less effective both for percentage of shoots rooting and number of roots per shoot. IBA at 3 mg/l resulted in 58% rooting and 4.8 roots per shoot. The shoots were treated with IBA for 12 days, then transferred to basal medium without hormones. Roots emerged after 3 to 4 weeks. A second rooting procedure was developed that involved dipping the shoots in IBA (1 mg/l) for 2 min and transferring them to auxin-free media. Ninty percent of the shoots rooted. One month after the shoots were rooted, plantlets were transferred to pots containing equal amounts of soil and perlite.

Chevre et al. (14) presented a detailed study of some of the requirements for culture initiation, multiplication, and rooting in juvenile and adult explant material of C. sativa. Axillary buds taken from 3-month-old seedlings were established and induced to form multiple shoots in media containing 1 mg/l BA. Further shoot multiplication resulted from subculturing in medium containing BA (1 mg/l) alone or in combination with IBA (0.1 mg/l). In one experiment, a multiplication rate of 30 was attained over a 10-week period. Adult explant material obtained from 5-year-old trees also produced multiple shoots when treated with BA at 1 mg/l. The adult material did not elongate in the multiplication medium as did the shoots produced by juvenile buds. A separate elongation medium was devised for the adult-derived shoots containing IAA (1 mg/l), adenine (1 mg/l), and 0.5% activated charcoal. Transfer of the adult shoots to this medium stopped multiplication and allowed shoot elongation. The authors report some limited success in rooting juvenile shoots by dipping them in a solution of IBA (1 g/l) for 20 s and culturing them on basal medium. Rooting was reported after 1 month. They were unable to root leafy shoots derived from adult explant material.

The in vitro shoot formation and plantlet regeneration of mature C. sativa has been reported by Biondi et al. (7) from bud cultures and by Vieitez et al. (79) from shoot tip explants of C. sativa x crenata obtained from winter-dormant stump sprouts of 10- to 18-year-old trees after cutting, cold treating, and sprouting. The adult hybrid trees were selected for their resistance to root rot. Shoot tip explants obtained as described above were soaked in sterile distilled water for 2 to 3 h in order to reduce tannin exudation. Explants treated with a range of BA concentrations (0.1, 0.5, and 1.0 mg/l) produced shoots measuring 1.0 to 2.5 cm with small axillary shoots sprouting from the shoot base after 3 to 4 weeks. The various BA treatments did not produce significantly different shoot growth. In fact, the zero concentration controls also produced shoots, though without axillary bud development. Various combinations of BA and NAA were tried in order to maximize shoot multiplication. The maximum rate of shoot multiplication was attained using 0.2 mg/l BA (4.3 shoots/culture). NAA by itself was found to neither produce multiplication nor elongation of the shoots. When used in combination with BA (0.2 mg/l) at 0.1 mg/l, NAA depressed the multiplication rate. Rooting of the shoots produced in the multiplication stage was accomplished by dipping the basal ends of the shoots in 0.5 or 1.0 g/l IBA for 15 or 2 min, respectively, then transferring the shoots to auxin-free agar medium. Rooting rates of 40% were attained using the 1.0 g/l, 2-min dip treatment.

San Jose and Vieitez (56) reported the production of adventitious buds and roots from epicotyl and hypocotyl sections, respectively, obtained from in vitro cultured embryonic axes of C. sativa. Adventitious roots were produced on hypocotyl sections treated with 0.1 mg/l NAA or IBA. Adventitious buds appeared on epicotyl sections when BA was present. A more detailed study of a similar nature was presented by San Jose et al. (57). Initial explants of C. sativa were obtained from plants grown in vitro. Adventitious buds were induced on epicotyl sections in the presence of BA (2 mg/l) plus IBA or NAA at 0.01 mg/l. Two to six buds were produced per explant after 2 to 4 weeks in culture. Shoots produced in these cultures were multiplied in medium containing 0.1 mg/l BA. Subcultures were made every 4 weeks using shoots developed from axillary buds. Cultures producing short shoots (i.e., 10 to 15 mm) were elongated by subculturing in medium containing 0.01 mg/l BA. Elongated shoots were then rooted in medium containing 2 mg/l IBA. Shoots were transferred to basal medium without hormones after 6 days. This procedure resulted in 77% rooting of the shoots. An alternate method of rooting shoots by dipping their basal end in a BA (0.5 or 1.0 g/l for 15 or 2 min) solution resulted in 100% rooting in some of the clones tested. Additional experiments resulting in adventitious root production in hypocotyl segments were also discussed.

Somatic embryogenesis, producing classic globular, heart-shaped, and torpedo-shaped embryoids, has been documented (24) in cotyledon cultures of C. sativa x crenata, a natural hybrid. The authors tested many combinations of 2,4-dichlorophenoxyacetic acid (2,4-D) and various cytokinins. Somatic embryogenesis took place in most of the cultures that contained 2,4-D alone or in combination with kinetin, BA, or zeatin. When the embryogenic callus was transferred to auxin-free medium, the embryoids failed to develop into plantlets.

Apical and lateral buds of C. dentata (52) obtained from 25-year-old, winter-dormant trees were treated with a forcing solution (8-hydroxyquinoline citrate) (53) to produce shoot extension on the cuttings. The bud explants obtained in this fashion were cultured in medium containing 0.5 mg/l BA. Proliferated shoots obtained from these cultures (5 to 10 shoots/culture) were dipped for 1 s in 3000 mg/l IBA dissolved in 70% alcohol. The shoots were then either transferred to basal medium containing activated charcoal or directly to flats containing sterile sand. Rooting occurred in both treatment regimes within 30 days. Rates of rooting were as high as 80%, but averaged 40 to 50%. The plantlets were successfully transferred to soil under greenhouse conditions.

A final paper concerning C. sativa is of interest because clonal propagation was accomplished without the use of plant growth regulators (68). Rescued embryos were allowed to develop on basal medium for 4 weeks. Each plantlet produced one apical and two nodal explants that were transferred to another basal medium of the authors own design, without hormones, where they developed vigorous tops and roots.

3. JUGLANS

Results of *in vitro* studies with *Juglans* are summarized in Table 2. An early report of successful aseptic culture of *Juglans* tissue resulted from research designed to develop a system for *in vitro* root initiation (15). Stem segments taken from 2-year-old seedlings of *J. nigra* were sterilized in a 1.5% NaOCl solution for 10 min, after which the bark was stripped from the shoots. The remaining stem cylinder was cut into 1-cm pieces for use as explant material. The explants were cultured in a basal medium containing NAA at 1.0 or 5.0 ppm and kinetin at 0.2 or 1.0 ppm. The explants were incubated in the dark at 23C. Well over half were lost to fungal contamination, but seven of the explants produced callus. BA at 0.2 ppm in combination with NAA at 5 ppm produced the most extensive callus growth. Callus seemed to be initiated in the cut areas near vascular traces. After 15 weeks, the cultures were cut into 0.5-cm pieces and transferred either to hormone-free media or to media containing kinetin (1 ppm) and NAA (4 ppm). The cultures on the hormone-free medium browned and ceased growth within the first week. The callus on hormone medium ceased growth after 1 month. The authors concluded that the lack of organogenesis in their system may have resulted from growing the cultures in the dark, from the presence of endogenous inhibitors, or from failure to add needed "requisite substances" to the culture media.

The micropropagation of Black walnut (*J. nigra*) was accomplished from rescued embryos and seedling shoot tip explants by Heile-Sudholt *et al.* (27). Multiple shoots were produced using BA at 10 μM for embryos and 22.2 μM for shoot tips. Shoot production and elongation were enhanced when the explants were cultured on liquid rather than solid media. Shoots were dipped in 10 mM IBA for 15 s and placed into moist sterile vermiculite under high humidity conditions to induce rooting. This treatment caused approximately 30% of the shoots to produce roots. Thirty percent of the rooted shoots were successfully transferred to soil under greenhouse conditions.

Rodriguez (54), using cotyledons of mature seed of *J. regia*, reported callus production and root organogenesis. The cotyledons were soaked in running tap water to remove "interfering phenolic compounds" then surface sterilized in a series of ethanol and sodium hypochlorite treatments. The cotyledons were first placed on plain mineral medium for 6 to 8 days for "preconditioning," then transferred to media containing various concentrations of auxins (IBA, 2,4-D, or NAA at 0, 0.4, 4, and 40 μM) and cytokinins (kinetin or BA at 0, 2, and 6 μM). The cultures were incubated under an 18-h photoperiod at 25C. No callus appeared in the zero controls. The callus induced by 2,4-D (4 and 40 μM) plus kinetin (2 and 6 μM) became partially habituated and no longer required auxins for growth. In addition, phenolic production increased, and after a time, these lines browned and died. The callus induced by IBA or NAA (4, 40 μM) alone or in combination with kinetin (2, 6 μM) was green in color and nodular. After 2 weeks, these cultures produced large amounts of anthocyanins that accumulated on the surface of the callus in "granular formations." Roots that were produced in these cultures appeared to originate from these granular formations and often contained anthocyanins. Interestingly, the authors found that although cytokinins were not required for root formation, their presence enhanced the auxin rooting effect.

TABLE 2. *Juglans*

Species	Maturity	Primary Explant	Result	Growth Regulator and Effective Concentration	Successful Transfer to Soil	Reference
nigra	J	stem segment from 2-year-old seedlings	callus	NAA (5 mg/l) + kinetin (0.2 mg/l)	--	Cummins and Ashby, 1969
nigra	J	apical buds	axillary branching root formation (?)	BA (2, 5 mg/l)	--	Heile *et al*., 1984
nigra	J	embryonic axes	multiple shoots	BA (10 μM)		
		shoot tips	multiple shoots	BA (22.2 μM)		
			root initiation	IBA (10 mM) (15-sec. dip)	yes	Heile-Sudhott *et al*., 1986
regia	J	cotyledons of mature seed	callus	2,4-D (4 to 40 μM) and kinetin (2 to 6 μM)	--	Rodriguez, 1982
			callus and root initiation	IBA or NAA (4, 40 μM) alone or in combination with kinetin (2, 6 μM)		
regia	J	immature and mature endosperm	roots	--	--	Cheema and Mehra, 1982
			callus	various auxins		
			callus with initiation of bud- and root-like structures	2,4-D (1 mg/l) + kinetin (1 mg/l)		
hindsii x *regia*	J	apical and lateral buds of seedlings	multiple shoots	BA (4.5 μM) + IBA (5 nM)		Driver and Kuniyuki, 1984
			root initiation (*in vitro*)	IBA (30 μM)		
			(direct to soil)	IBA (5 mM) dip	yes	
regia	J	immature cotyledons	embryogenesis	BA (1 mg/l) + IBA (0.01 mg/l) + kinetin (2.0 mg/l)	yes	Tulecke and McGranahan, 1985
hindsii	J	cotyledons	embryogenesis	as for *regia*	--	

Immature and mature endosperm of *J. regia* was cultured on an array of auxins and cytokinins and gibberellic acid (GA_3) to determine the effect of embryonic tissue and/or an embryo factor (GA_3) for induction of embryoids (13). Roots were obtained via organogenesis in immature endosperm cultures, and shoot bud-like structures were produced in cultures containing NAA and kinetin at 0.2 mg/l. Mature endosperm tissue also produced roots and bud-like structures when 2,4-D (1 mg/l) and kinetin (1 mg/l) were present in the culture medium. The initiation and subsequent development of these various structures occurred in the absence of embryonic tissue or GA_3. Interpretation of these results is somewhat complicated by the fact that the media also contained "complex growth substances" such as casein hydrolysate, yeast, and malt extract singly and in combination.

The proliferation of apical and lateral buds of *J. hindsii x regia* obtained from seedlings was reported by Driver and Kuniyuki (18) in 1984. Nodal explants were surface sterilized in 0.26% (v/v) NaOCl for 20 min, rinsed three times in distilled water, and placed on basal media without hormones for 1 week under constant light. Injured or contaminated cultures were eliminated during this time period. Explants were then transferred to media containing BA (0.45 to 45 μM) alone or in combination with IBA (5 nM to 5 μM). BA concentrations of less than 4.5 μM produced one shoot per node; concentrations greater than 4.5 μM resulted in three shoots per node, accompanied by excessive callus production. The addition of IBA to 5 μM increased shoot production; higher levels produced increasing amounts of callus. Optimal shoot production occurred at 4.5 μM BA and 5 nM IBA. Rooting of the microshoots was accomplished both *in vitro* and in direct-to-soil systems. *In vitro* rooting tests were run using IBA and NAA at concentrations from 0.1 to 50 μM. IBA at 30 μM induced 70% rooting. At less than 30 μM IBA, no rooting occurred; at greater than 30 μM, excessive callus was produced. Plantlets produced under these conditions were subsequently transferred to soil after suitable acclimation. Interestingly, NAA (3.0 μM) induced 80% rooting of the microshoots, but these plantlets could not be hardened off to *ex vitro* conditions. Shoots were rooted under nonsterile conditions using a 5 mM IBA dip, followed by direct planting in a mixture of sand:peat:perlite (1:1:1). Because repeated subculturing of the shoots produced lower multiplication rates and greater callusing, a specialized basal medium was produced that sustained shoot production and increased the shoot growth rate four fold.

Heile *et al*. (26) have reported shoot proliferation in *J. nigra* using shoot tip cultures of greenhouse-grown seedlings. Optimal concentrations of BA for culture establishment were not reported, but 2 and 5 ppm produced the most branching, while higher levels induced excessive callusing. Browning of cultures was avoided by subculturing at 1, 3, and 5 days after culture initiation and every 14 days thereafter. The authors reported that rooting of the microshoots was difficult; no data were given.

The induction of somatic embryogenesis in immature cotyledonary tissue of *J. regia* with the development of plants from the asexually produced embryos (74) represents a major advance in the clonal propagation of the genus, although the extent of genetic variability present has not yet been determined. Immature cotyledons collected 6 to 11 weeks after pistillate bloom were placed on a conditioning medium containing BA (1 mg/l), kinetin (2 mg/l), IBA (0.01 mg/l), and the amino acid glutamine (250 mg/l). After 2 to 4 weeks, the explants were transferred to basal medium without hormones or glutamine. Transfers were made at 2- to 4-week intervals thereafter. Somatic embryos appeared from 6 weeks to 6 months after initial culturing. Embryos that developed from the cotyledons were removed and subcultured on basal medium. Mature somatic embryos were found to require an 8- to 10-week cold treatment at 2 to 4C in order to be able to proceed with a normal developmental pattern (primary root and shoot development, etc.). Embryogenic cultures have been maintained for more than a year and a half, and more than 40 plants have been successfully transferred to soil.

4. QUERCUS

Jacquiot (32, 33) reported callus production from in vitro cultured cambial tissues of Q. sessiliflora and suber. In another study, internodal stem segments of 2- to 4-month-old Q. rubra seedlings were placed into sterile culture media containing various combinations of BA (0, 0.01, 0.1, 1, or 5 mg/l) and NAA (0, 0.1, 1, 3, 5, or 10 mg/l) (59). Callus was produced from some of the cultures (Table 3) and could be sustained if subcultured. Callus cultures produced from 0.1 mg/l BA in combination with 5 mg/l NAA produced morphologically anomalous structures termed organoids. Rapid multiplication of these structures was reported when both solid and liquid media were supplemented with various combinations of BA and NAA. Adventitious roots were produced by organoids grown on media containing auxins (1 to 5 mg/l) with or without cytokinin. Microscopic observations revealed roots with normal morphology and vascular connection to the vascular strands of the organoid. The production of these aberrant structures was thought to be, at least in part, the product of the intervening callus stage.

The production of nodular structures on the surface of callus tissue was noted by Linberger (38) in in vitro cultures of Q. palustris. Nodal segments of rapidly elongating stem tips were cultured on medium containing NAA. Histological examination of these structures revealed a well-defined epidermal layer, cellular organization, and differentiation of tracheary elements. No root or shoot differentiation was observed.

Plantlet formation was reported using explants of axillary buds by Pardos (49) in Q. suber and by Chalupa (12) in Q. robur. Srivastava and Steinhauer (66) used mature embryos of Q. lebani and an array of complex hormonal supplements and combinations on both solid and liquid media. Depending upon the combination of explant source, hormonal treatment, and media type, the authors were able to produce callusing, shoot and root differentiation, and embryoids (see Table 3 for examples of growth regulators and concentrations).

Rapid multiplication of Q. robur axillary buds obtained from 2-month-old to 1-year-old seedlings was induced by BA at concentrations ranging from 0.2 to 1.0 mg/l (11). Both BA and kinetin were tested over a range of concentrations (0.2 to 5.0 mg/l) for their ability to stimulate shoot growth and multiplication. Kinetin proved to be less effective at all concentrations tested. Relatively low levels of BA (0.2 to 0.8 mg/l) were effective in stimulating the multiplication and elongation of the axillary buds. BA at 2.0 to 5.0 mg/l produced shoots that did not elongate. The growth and proliferation of the shoots was not affected by the addition of NAA or IBA at 0.05 to 0.1 mg/l. Shoots 2 to 4 cm in length, excised from proliferating cultures and cultured in agar media containing both IBA (0.3 mg/l) and NAA (0.1 mg/l), rooted at rates of 70 to 95%. Plantlets were successfully transferred to soil with a 90 to 95% survival rate.

TABLE 3. Quercus

Species	Maturity	Primary Explant	Result	Growth Regulator and Effective Concentration	Successful Transfer to Soil	Reference
rubra	J	internodal stem segment (2- to 4-month-old seedlings)	callus and "organoids"	BA (5 mg/1) + NAA (0.1 mg/1)	--	Seckinger et al., 1979
			roots (from organoids)	NAA (1 to 5 mg/1)		
palustris	--	stem tips	callus and "nodular" structures	NAA (-)	--	Lineberger, 1980
suber	J	axillary buds	axillary bud elongation	BA and NAA (0.5 μM)	--	Pardos, 1981
			root initiation	(-)		
lebani	J	embryos	multiple shoot buds and roots	2iP (5 mg/1) + IAA (2 mg/1)	--	Srivastava and Steinhauer, 1982
		embryo segments				
		-radicular end	roots and callus	2iP (5 mg/1) + IAA (2 mg/1)		
		-cotyledon end	shoots	2iP (5 mg/1) + IAA (2 mg/1)		
		-middle segment	callus, shoot buds, and roots	2iP (5 mg/1) + IAA (2 mg/1)		
robur	J	shoot tips and nodal segments	axillary shoot multiplication	BA (0.2 to 1.0 mg/1)		Chalupa, 1984
			root initiation	IBA and NAA (0.3 mg/1)	yes	
robur	J	embryos and apical buds (3 to 4-month-old seedlings)	shoot development	BA (1 mg/1)	--	Vieitez et al., 1985
			shoot multiplication	BA (0.1 mg/1)		
			root initiation	IBA (0.5 g/1 or 1.0 g/1) (10- or 2-min. dip)		
	M	apical buds from stump sprouts of 50-year-old trees	shoot development	BA (1 mg/1)		
			shoot multiplication	BA (0.1 mg/1)		
			root initiation	IBA (0.5 g/1 or 1.0 g/1) (10- or 2-min. dip)		
robur	J	apical and axillary buds (10 to 20-day-old to 2-year-old seedlings)	shoot elongation	BA (2.2 μM), IBA (2.5 μM), GA_3 (0.3 μM)	--	Pevalek-Kozlina and Jelaska, 1986
			adventitious bud production	BA (0.1 μM), kinetin (0.5 μM), 2iP (0.5 μM)		
petraea	J	as above	as above			
			adventitious root initiation	BA (1.3 μM), kinetin (0.9 μM), IBA (4.3 μM)	--	
shumardii	J	nodal stem sections	multiple shoot proliferation	BA (2 mg/1)		Bennett and Davies, 1986
			root initiation	IBA (0.5 g/1) (15-min. dip)	yes	

Vieitez et al. (80) presented evidence for plantlet regeneration from both juvenile and mature explants of Q. robur. The authors established cultures from embryos, apical buds of 3- to 4-month-old seedlings, and apical buds of stump sprouts taken from 50-year-old trees. Initial shoot development, shoot multiplication, and root initiation were induced in all of the explant types by similar hormonal treatments. Cultures were initiated on medium containing 1 mg/l of BA. Shoot multiplication was encouraged by transferring the shoots to a medium containing a lower level of BA (0.1 mg/l). Shoots 2 cm or more long taken from cultures of juvenile or mature origin were dipped in an IBA solution and placed in an auxin-free medium. Root primordia appeared in 15 days in the juvenile material and 20 days in the mature material. A 2-min dip in a 1 g/l IBA solution produced 83% rooting of the juvenile material and 63% rooting of the shoots derived from mature trees.

Pevalek-Kozlina and Jeleska (50) worked with an interesting array of juvenile material of two Quercus species (i.e., robur and petraea). Shoot apices from 10- and 20-day-old seedlings and apical and axillary buds from plants harvested at 2- to 4-month intervals to 2 years of age were used as explants. Bud cultures of both species were established in a medium containing BA (2.2 µM), IBA (2.5 µM), and GA_3 (0.3 µM). Explant age interacted in a complex manner with both basal medium and hormonal treatment. Elongation and adventitious bud development were obtained for both species. Root development from callus that developed at the basal end of the explant was reported for Q. petraea.

Plantlet regeneration in Q. shumardii (6) was accomplished using nodal stem sections of etiolated and light-grown seedling stock plants 3 to 36 months old. The nodal sections were put through an extensive four-stage surface sterilization process involving commercial sterilants and two separate ethanol dips. The explants were cultured in liquid basal media containing BA or 2iP at 0, 0.5, 1, 2, 3, or 5 mg/l. The cultures were grown in the light under a 16-hour photoperiod. The explants were transferred daily for 5 days and then at 7-day intervals. 2iP had no effect on shoot multiplication. Multiple shoots were produced at concentrations of BA >1 mg/l. To try to optimize both shoot multiplication and growth in a single subculturing medium, a series of hormonal concentrations and combinations were tested. The hormonal mix settled upon was BA (2 mg/l), IBA (0.5 mg/l), and GA_3 (1 mg/l). Shoots treated with this mix could be subcultured three times, giving an average of six shoots per transfer, two of which averaged >10 mm in length. Shoots >10 mm in length were selected for rooting experiments. The basal ends of the shoots were dipped in IBA (0.5 g/l) for 0, 5, 10, 15, and 30 min, placed in Jiffy-7™ pellets on petri dishes, and covered with inverted cups. The shoots were then incubated in the dark for 7 days and returned to a lighted growth chamber. The 15-min IBA dip treatment was optimal and produced 73% rooting. The plantlets were successfully transferred to soil and then to greenhouse conditions.

Schwarz and Schlarbaum (unpublished) used Q. alba and rubra explants obtained from rescued embryos and apical and axial buds (taken from 3-month-old seedlings) to develop in vitro methodologies for the micropropagation of mature oaks. Because of the known recalcitrance of mature oaks to being vegetatively propagated by conventional means (22, 62), micropropagation was approached through the utilization of explant material of widely differing states of development. Maximal multiple shoot production with respect to number of explants with multiple shoots and number of shoots per explant occurred at 4.44 µM BA. The addition of

NAA to this treatment level of BA reduced both parameters in embryo explant cultures but enhanced both when it was applied to explants obtained from 3-month-old seedlings. Shoot development occurred in 100% of the seedling explants treated with 4.44 µM BA and 1 nM NAA. This treatment produced more shoots per explant and also produced a number of shoots with expanded leaves. It is interesting that the most effective level of BA for multiple shoot production for both rescued embryos and seedling apical buds is 4.44 µM. However in one system, the addition of NAA induced 100% multiple bud development, and in the other, this response was severely inhibited, even at 1 nM.

5. DISCUSSION

Several interpretive approaches are possible with respect to the literature reviewed for this chapter. The two most inviting are from the points of view of mechanism and manipulation. Minocha (46), in a recent review, chose a mechanistic view to try and relate the role of growth regulators and morphogenesis to various avenues of the *in vitro* culture of forest trees. Minocha listed five factors inherent in much of the literature that interfered with a concise synthesis and critical assessment of the role of plant growth regulators in cell and tissue culture of woody plants. These factors are:

1. There is an abundance of qualitative observations but little or no quantitative data for the induction of callus, regeneration of roots and shoots, and various aspects of micropropagation from organized tissues.
2. Interpretation of the results of experiments with reference to cause and effect is difficult because of the long tissue incubation times necessary before a response is manifest (i.e., the composition of the media can change significantly).
3. The "physiological status" of the various explant tissues or calli is not known and is not uniform between experiments, which decreases reproducibility.
4. Uptake and possible metabolism of the regulator is usually not followed.
5. ".... we do not understand the biochemical basis of action of any commonly used growth regulator even in the model experimental systems."

The first four factors are, to a great extent, reparable through attention to experimental design. Overcoming the fifth factor is indeed complex, but its mastery may be in part facilitated by an improvement of the first four.

With these strictures in mind, I chose to review the literature dealing with three genera (*Castanea*, *Juglans*, and *Quercus*) from a manipulative point of view. The growth regulator applications and cultural manipulations reviewed here are typical of those utilized for clonal micropropagation in other hardwood genera (see reviews 3, 9, 19, 20, 64, 65, 70, 81). Tables 1, 2, and 3 provide a summary of the source of primary explant tissue and the growth regulator and concentration applied that produced the best results under the specific set of cultural conditions used by each investigator. Although not discussed here, there seems to be a definite media effect/interaction with the other culture variables with respect to multiple shoot production and rooting (18, 35, 66, 78, 79, 80).

The "state-of-the-art" in each of the three genera (*Castanea*, *Juglans*, and *Quercus*) has developed rapidly with respect to the utilization of embryonic and seedling explant material. In a recent review of hardwoods and their regeneration from tissue and organ culture, Sommer and Wetzstein (65) concluded that "...it is obvious that the culture types for propagation of hardwood trees are not much different from those used for other plant species. The fundamental difference is in most cases only juvenile tissues have been used or given successful results." This conclusion and observation ring true with respect to the three genera reviewed here; however, the frequency of plantlet regeneration using mature explant material is increasing (32, 52, 80).

The growth regulator of choice used for the initial explant isolation (stage 1) to promote embryo or bud elongation and, in some cases, multiple shoot proliferation was BA at concentrations from 0.5 to 5.0 mg/l. The effectiveness of BA in promoting in vitro multiple shoot formation in forest trees is well documented in the literature. Zaerr and Mapes (82) list examples of some 25 gymnosperm and 18 angiosperm species and varieties that have responded to the in vitro application of BA by producing multiple shoots. The utilization of BA by various investigators, rather than the use of one of the naturally occurring cytokinins (e.g., zeatin or 2iP), may be in part a function of cost. More likely, BA is used because it is consistently more effective in promoting the desired result--the generation of multiple shoots (82). The precise reason for BA's greater efficacy is not known; however because of its artificial nature, BA might not be readily metabolized by plant tissues. BA's tissue persistence/titer might then be greater than the naturally occurring growth regulators, which might allow for its increased effectiveness. Interestingly, Hu and Wang (30) found that BA was the most frequently used cytokinin in culture establishment and proliferation of meristem, shoot tip, and bud cultures. As mentioned above, the effective levels used in culture establishment (i.e., stage 1) were found to vary from 0.5 mg/l in winter-dormant apical and lateral buds of C. dentata (52) to 5.0 mg/l in seedling shoot tips of C. sativa (55). Closer inspection of Tables 1 through 3 reveals essentially no relationship between the age of the explant (i.e., developmental maturity) and the concentration of growth regulator used to establish the stage 1 cultures. What seems to be of primary importance in these and other forest tree genera may not be the concentration of the growth regulator but the physiological or developmental state of the explant at the time of initial exposure (8, 52, 79, 80). The importance of the type of explant is underscored by the work of Srivastava and Steinhauer (66). Using a fixed growth regulator treatment and different segments of Q. lebani embryos, they were able to document differing morphogenetic response.

The initial culture phase (stage 1) was often followed by a transfer to media containing five- to tenfold less cytokinin. This decrease in cytokinin concentration stimulated and/or allowed shoot proliferation and, in many cases, elongation. BA was again the growth regulator of choice in these three genera for the shoot multiplication stage (stage 2). BA was occasionally used in combination with one or two other cytokinins (50) and various auxins (14, 18, 55, 56, 78). There were several reports of stage 2 activity taking place as a result of the stage 1 isolation and growth regulator exposure (6, 11, 18, 35, 42, 50, 56). However in most cases, the desired level of shoot multiplication was achieved after a transfer to lower levels of the growth regulator or regulators.

The final morphogenetic event in plantlet regeneration is the production of adventitious roots on the proliferated shoots. Successful rooting, either *in vitro* or directly in soil, usually followed a period of elongation of the newly produced shoots (55, 77, 78). The elongation of the proliferated shoots was encouraged by transferring the cultures to low levels of cytokinin (i.e., concentrations well below levels used to enhance multiple shoot production) or to medium containing no cytokinins. Two auxins (IBA and NAA) were used routinely for the purpose of root initiation (see Tables 1, 2, and 3). Both IBA and NAA are synthetic analogs of IAA. Their use in various aspects of plant cell and tissue culture is widespread (82). The reason for their sometimes increased effectiveness over IAA is unknown. However, reasons suggested in the literature range from a suspected longer tissue half-life because of their synthetic nature (82) to the ability of IBA to stimulate less ethylene (an inhibitor of rooting in some plants) than is stimulated by IAA (31). The auxins were added to the basal medium or applied as a dip to shoots placed directly in soil for various incubation times that depended upon concentration. In at least one case (78), the rooting hormone was found to inhibit the number and length of roots developed if it was left in contact with the shoot for longer than 12 days. In this case, the shoots were transferred to auxin-free medium at 12 days in order to avoid the inhibiting effect.

In vitro somatic embryogenesis has been successful in a number of tropical and temperate woody tree species (4, 39, 48, 65, 72). Two examples of somatic embryogenesis are reviewed here, one each in *Castanea* (24) and *Juglans* (74). Both involved the use of cotyledonary tissue and the presence of an auxin in the culture medium used in the initial stage of culture. This is a typical requirement for the initiation of somatic embryo development (4). It is very encouraging to find that in one of these systems plantlets were regenerated and successfully transferred to soil (74). In each case, the explants were exposed to a conditioning medium that contained the single type or mixture of growth regulators found to initiate the appropriate morphogenetic responses necessary to produce somatic embryos. Although this initial growth regulator treatment is not always required (23), its efficacy has been reaffirmed in the production of somatic embryos in pecan (44), sweetgum (63), and yellow poplar (43). The production of fully developed somatic embryos usually involved the eventual transfer of the cultures to a hormone-free basal medium. The removal of growth regulators from cultures to allow either induction of embryoids or their further development has been reported for a wide variety of embryogenic systems (51, 60).

6. SUMMARY

The approach to developing cultural procedures for the *in vitro* production of forest trees is for the most part a directed process of elimination. An experimental approach governed by a clear picture of the mechanism of growth regulator action is not yet possible. However, the literature reviewed for *Castanea*, *Juglans*, and *Quercus* illustrates that progress is being made towards the establishment of *in vitro* propagative systems for forest trees through the application of a variety of manipulative approaches. These approaches involved the use of growth regulators and a complex set of culture variables.

ACKNOWLEDGEMENT

The author wishes to acknowledge the support of the USDA Forest Service Southern Region: Region 8.

REFERENCES

1. Abbott AJ 1977 Propagating temperate woody species in tissue culture. Sci Hort 28:155-162
2. Abbott AJ 1978 Practice and promise of micropropagation of woody species. Acta Horticulturae 79:113-127
3. Ahuja, MR and Muhs HJ 1985 In vitro techniques in clonal propagation of forest tree species. In: Schafer-Menuhr A, ed. In Vitro Techniques Propagation and Long Term Storage. Boston: Martinus Nijhoff/Dr. W. Junk Publishers
4. Ammirato PV 1983 Embryogenesis. In: Evans DA, Sharp WR, Ammirato PVA, and Yamada Y, eds. Handbook of Plant Cell Culture. Vol. I. Techniques for Propagation and Breeding, pp 82-123. New York: Macmillan Publishing Co.
5. Bayliss MW 1973 Origin of chromosome number variation in cultured plant cells. Nature 246:529-530
6. Bennett LK and Davies FT 1984 In vitro propagation of Quercus shumardii Buckl. HortScience 19:576
7. Biondi S, Canciani L, DePaoli G, and Bagni N 1981 Shoot formation from bud culture of mature chestnut. In: AFOCEL Association, ed. Proceedings of Colloque International sur la culture in vitro des essences forestiers, pp 180-185. Fontainebleau: IUFRO
8. Bonga JM 1987 Clonal propagation of mature trees: problems and possible solutions. In: Bonga JM and Durzan DJ, eds. Cell and Tissue Culture in Forestry, pp 249-271. Boston: Martinus Nijoff Publishers
9. Bonga JM and Durzan DJ, eds 1982 Tissue Culture in Forestry. Boston: Martinus Nijhoff/Dr. W. Junk, Publishers
10. Brown CL and Sommer HE 1982 Vegetative propagation of dicotyledonous trees. In: Bonga JM and Durzan DJ, eds. Tissue Culture in Forestry, pp 109-149. Boston: Martinus Nijhoff/Dr. W. Junk Publishers
11. Chalupa V 1984 In vitro propagation of oak (Quercus robur L.) and Linden (Tilia cordata Mill.). Biologia Plantarum (Praha) 26:374-377
12. Chalupa V 1979 In vitro propagation of some broad-leaved forest trees. Commun Inst Forest Cechosl 11:159-170
13. Cheema GS and Mehra PN 1982 Morphogenesis in endosperm cultures. In: Fujiwara A, ed. Proceedings of the 5th International Congress of Plant Tissue and Cell Culture, pp 111-112. Tokyo: The Japanese Association for Plant Tissue Culture
14. Chevre AM, Gill SS, Mouras A, and Salesses G 1983 In vitro vegetative multiplication of chestnut. Journal of Horticultural Science 58:23-29
15. Cummins N and Ashby WC 1969 Aseptic culture of Juglans nigra L. stem tissues. Forest Sci 15:102-103
16. D'Amato F 1977 Cytogenetics of differentiation in tissue and cell cultures. In: Reinert J and Bajaj YPS, eds. Plant Cell, Tissue and Organ Culture, pp 343-357, Berlin: Springer-Verlag
17. Douglas GC 1985 Control and utilization of adventitious bud formation in Populas explants. In: Schafer-Menuhr A, ed. In Vitro Techniques Propagation and Long Term Storage, pp 53-58. Boston; Martinus Nijhoff/Dr. W. Junk Publishers
18. Driver JA and Kuniyuki AH 1984 In vitro propagation of Paradox walnut rootstock. HortScience 19:507-509

19. Durzan DJ 1985 Tissue culture and improvement of woody perennials: An overview. In: Henke RR, Hughes KW, Constantin MJ, and Hollaender A, eds. Tissue Culture in Forestry and Agriculture, pp 233-256. New York: Plenum Press
20. Eriksson T, Fridborg G, and Von Arnold S 1977 Tissue and cell culture of forest trees as a tool of vegetative propagation. In: Vegetative Propagation of Forest Trees - Physiology and Practice. Lectures from a Symposium in Uppsala, Sweden, February 16-17, 1977, pp 17-26. Sweden: The Swedish University of Agricultural Sciences
21. Farnum P, Timmis R, and Kulp JL 1983 Biotechnology of forest yield. Science 219:694-702
22. Flemer W III 1962 The vegetative propagation of oaks. Proc Plant Prop Soc 12:168-171
23. Gharyal PK and Maheshwari SC 1981 In vitro differentiation of somatic embryoids in a leguminous tree - Albizzia lebbeck L. Naturwissenschaften 68:379-380
24. Gonzales ML, Vieitez AM, and Vieitez E 1985 Somatic embryogenesis from chestnut cotyledon tissue cultured in vitro. Scientia Horticulturae 27:97-103
25. Haissig BE 1965 Organ formation in vitro as applicable to forest tree propagation. Botanical Review 30:607-626
26. Heile C, Gaffney G, Preece J, Meyers O, and Van Sanbeek J 1984 In vitro culture of Juglans nigra seedling shoot tips. HortScience 19:100
27. Heile-Sudholt C, Huetteman CA, Preece JE, Van Sambeek JW, and Gaffney GR 1986 In vitro embryonic axis and seedling shoot tip culture of Juglans nigra L. Plant Cell, Tissue and Organ Culture 6:189-197
28. Hicks GS 1980 Patterns of organ development in plant tissue culture and the problem of organ determination. The Botanical Review 46:1-23
29. Hu C-Y and Scrivani EC 1977 Effect of kinetin on callus morphogenesis from stem apex explants of American chestnut (Castanes dentata) winter buds. Plant Phys 59:2
30. Hu C-Y and Wang PJ 1983 Meristem, shoot tip, and bud culture. In: Evans DA, Sharp WR, Ammirato PV, and Yamada Y, eds. Handbook of Plant Cell Culture. Vol I. Techniques for Propagation and Breeding, pp 177-227. New York: Macmillan Publishing Company
31. Jacobs WP 1979 Plant Hormones and Plant Development. New York: Cambridge University Press
32. Jacquiot C 1952 Sur les phenomenes d'histogenese observes dan des cultures in vitro de tissu cambial de chenes (Quercus sessiliflora Sm., Q. pendunculata Ehr., Q. suber L.). Comptes Rendus Hebdomadaires des Siances de l'Academic des Sciences 234:1468-1470
33. Jacquiot C 1956 Sur les besoins en auxine et les caracteres morphologiques externes des cultures de tissu cambial de quelques especes d'arbres. Comptes Rendus Hebdomadaires des Seances de l'Academic des Sciences 243:510-512
34. Karnosky DF 1981 Potential for forest tree improvement via tissue culture. BioScience 31:114-120
35. Keys RN and Cech FC 1981 Plantlet formation in American chestnut embryonic tissue in vitro. In: Proceedings of the 2nd North Central Tree Improvement Conference, pp 189-194
36. Kleinshmit J 1974 A programme for large scale cutting propagation of Norway spruce. JZJ For Sci 4:359-366

37. Lester DT and Berbee JG 1977 Within-clone variation among black poplar trees derived from callus culture. For Sci 23:122-131
38. Lineberger D 1980 Anatomy of organized structures produced on sovereign pin oak callus in vitro. HortScience 15:437
39. Litz RE 1985 Somatic embryogenesis in tropical fruit trees. In: Henke RR. et al., eds. Tissue Culture in Forestry and Agriculture, pp 179-193. New York: Plenum
40. Maynard CA 1986 Population genetics of forest trees: Implications for the application of in vitro techniques. In Vitro Cellular and Developmental Biology 22:231-233
41. McCown BH 1986 Woody ornamentals, shade trees, and conifers. In: Zimmerman RH, et al., eds. Tissue Culture as a Plant Production System for Horticultural Crops, pp 333-342. Dordrecht: Martinus Nijhoff Publishers
42. McPheeters K, Skirvin RM, and Bly-Monnen CA 1980 Culture of chestnut (Castanea spp.) in vitro. HortScience 15:417-418
43. Merkel SA and Sommer HE 1986 Somatic embryogenesis in tissue culture of Liriodendron tulipifera. Can. J. For. Res. 16:420-422
44. Merkel SA, Wetstein HY, and Sommer HE 1987 Somatic embryogenesis in tissue cultures of pecan. HortScience 22(1):128-130
45. Miller LR and Murashige T 1976 Tissue culture propagation of tropical foliage plants. In Vitro 12:797-813
46. Minocha SC 1987 Plant growth regulators and morphogenesis in cell and tissue culture of forest trees. In: Bonga JM and Durzan DJ eds. Cell and Tissue Culture in Forestry Volume 1 General Principles and Biotechnology, pp 50-66. Boston: Martinus Nijhoff Publishers
47. Murashige T 1974 Plant propagation through tissue culture. Ann Rev Plant Physiol 25:135-166
48. Narayanaswamy S 1979 Regeneration of plants from tissue cultures. In: Reinert J and Baja YPS, eds. Plant, Cell, Tissue and Organ Culture, pp 179-248. New York: Springer - Verlag
49. Pardos JA 1981 In vitro plantlet formation from stem pieces of Quercus suber. In: AFOCEL, ed. Proceedings of Colloque International sur la Culture 'in vitro' des Essences Forestieres, pp 186-190. Fontainebleau: IUFRO
50. Pevalek-Kozlina B and Jelaska S 1986 In vitro growth and development of oaks (Quercus robur and Q. petraea). Acta Bot Croat (in press)
51. Radojevic L 1979 Somatic embryos and plantlets from callus cultures of Paulownia tomentosa Steud. Z. Pflanzenphysiol. 91:57-62
52. Read PE, Hosier MA, and Quiguang Y 1985 Tissue Culture of Chestnuts. Annual Meeting of the Northern Nut Growers Association Report
53. Read PE, Qiguang Y, Hosier MA, and Lee SW 1985 Forcing solutions: A novel plant growth regulator delivery method for in vitro culture of woody species. HortScience 19:592
54. Rodriquez R 1982 Callus initiation and root formation from in vitro culture of walnut cotyledons. HortScience 17:195-196
55. Rodriquez R 1982 In vitro propagation of Castania sativa Mill. through meristem-tip culture. HortScience 17:888-889
56. San Jose MC and Vieitez E 1984 Morphogenesis in sections of chestnut plantlets cultivated in vitro. Anales de edafologia y agrobiologia 43:587-598
57. San Jose MC, Vieitez AM, and Vieitez E 1984 In vitro plantlet regeneration from adventitious buds of chestnut. Journal of Horticultural Science 59:359-365

58. Savidge RA 1983 The role of plant hormones in higher plant cellular differentiation. I. A critique. Histochem J 15:437-445
59. Seckinger GR, McCown BH, Struckmeyer BE, and Durbin RD 1978 Production and rapid multiplication of organoids from callus cultures of Quercus rubra L. (Red Oak). HortScience 13:Abstract 135
60. Sharp WR, Sondahl MR, Caldas LS, and Maraffa SB 1980 The physiology of in vitro asexual embryogenesis. Hortic. Rev. 2:268-310
61. Sheridan WF 1975 Plant regeneration and chromosome stability in tissue cultures. In: Ledoux L, ed. Genetic Manipulations with Plant Material, pp 263-295. New York: Plenum Press
62. Skinner HT 1952 Vegetative propagation of oaks and suggested research techniques. Proc Plant Prop Soc 2:81-85
63. Sommer HE and Brown CL 1980 Embryogenesis in tissue cultures of sweetgum. Forest Sci. 26:257-260
64. Sommer HE and Caldas LS 1981 In vitro methods applied to forest trees. In: Plant Tissue Culture. Thorpe Edition, pp 349-358. New York: Academic Press, Inc.
65. Sommer HE and Wetzstein HY 1984 Hardwoods. In: Ammirato PV, Evans DA, Sharp WR, and Yamada Y, eds. Handbook of Plant Cell Culture. Crop Species, Volume 3, Chapter 19, pp 511-540. New York: Macmillan Publishing Company
66. Srivastava PS and Steinhauer A 1982 In vitro culture of embryo segments of Quercus libani: Organogenesis and callus growth as a different response to experimental conditions. Z Pflanzenphysiol Bd 106:93-96
67. Stone OM 1978 The production and propagation of disease free plants. In: Hughes KW, Henke RH, and Constantin M, eds. Propagation of Higher Plants Through Tissue Culture: A Bridge Between Research and Application, pp 25-34. Technical Information Center, United States Department of Energy. Available from NTIS, U.S. Department of Commerce, Springfield, Virginia
68. Strullu DG, Grellier B, Marciniak D, and Letouze R 1986 Micropropagation of chestnut and conditions of mycorrhizal syntheses in vitro. New Phytol 102:95-101
69. Styer DJ and Chin CK 1983 Meristem and shoot-tip culture for propagation, pathogen elimination, and germplasm preservation. In: Janick J, ed. Horticultural Reviews, Vol. 5, pp 221-277. Connecticut: AVI Publishing Company
70. Thompson DG 1984 Clonal reforestation: Forests of the future? In: Duryea ML and Brown GH, eds. Seedling Physiology and Reforestation Success, pp 3-28. Boston: Martinus Nijholl/Dr. W. Junk Publishers
71. Thorpe JA 1978 Physiological and biochemical aspects of organogenesis in vitro. In: Thorpe TA, ed. Frontiers of Plant Tissue Culture 1978, pp 47-58. Calgary: International Congress of Plant Tissue and Cell Culture
72. Tisserat B, Esan BB, and Murashige T 1979 Somatic embryogenesis in angiosperms Hortic Rev 1:1-78
73. Torrey J 1966 The initiation of organized development in plants. Adv Morphogenesis 5:39-91
74. Tulecke W and McGranahan G 1985 Somatic embryogenesis and plant regeneration from cotyledons of walnut, Juglans regia L. Plant Science 40:57-63
75. Vieitez E 1974 Vegetative propagation of chestnut. New Zealand Journal of Forestry Science 4:242-252

76. Vieitez AM and Vieitez E 1980a Plantlet formation from embryonic tissue of chestnut grown *in vitro*. Physiol Plant 50:127-130
77. Vieitez AM and Vieitez ML 1980b Culture of chestnut shoots from buds *in vitro*. Journal of Horticultural Science 55:83-84
78. Vieitez AM and Vieitez ML 1982/83 *Castanea sativa* plantlets proliferated from axillary buds cultivated *in vitro*. Scientia Horticulturae 18:343-351
79. Vieitez AM, Ballester A, Vieitez ML, and Vieitez E 1983 *In vitro* plantlet regeneration of mature chestnut. J Horticultural Science 58:457-463
80. Vieitez AM, San-Jose MC, and Vieitez E 1985 *In vitro* plantlet regeneration from juvenile and mature *Quercus robur* L. J Horticultural Science 60:99-106
81. Winton L and Huhtinen O 1976 Tissue culture of trees. In: Miksche JP, ed. Modern Methods in Forest Genetics, Chapter 12, pp 243-264. New York: Springer-Verlag
82. Zaerr JB and Mapes MO 1988 Action of growth regulators. In: Bonga JM and Durzan DJ, eds. Tissue Culture in Forestry, pp 231-255, Boston: Martinus Nijhoff/Dr. W. Junk Publishers
83. Zimmerman RH 1985 Application of tissue culture propagation to woody plants. In: Henke RH, Hughes KW, Constantin MJ, and Hollaender A, eds. Tissue Culture in Forestry and Agriculture, pp 165-177. New York: Plenum Press

Plant Growth Regulation 6: 137–169

7. THE ROLE OF PLANT GROWTH REGULATORS IN FOREST TREE CAMBIAL GROWTH

C.H.A. LITTLE AND R.A. SAVIDGE

Canadian Forestry Service - Maritimes, P.O. Box 4000, Fredericton, New Brunswick E3B 5P7, Canada, and Department of Forest Resources, University of New Brunswick, Fredericton, New Brunswick E3B 6C2, Canada

ABSTRACT
The regulation of cell-division activity in the vascular cambium and of secondary xylem and phloem development is reviewed for temperate-zone tree species in relation to auxins, gibberellins, abscisic acid, cytokinins, and ethylene. Representatives of the first four of these PGR classes (IAA, GA_1, GA_4, GA_7, GA_9, GA_{20}, ABA, Z, ZR, DCA) have been identified conclusively by mass spectrometry in the cambial region in some Pinaceae, but not in any hardwood species. Endogenous ethylene has yet to be definitively characterized in this region in any species. Evidence concerning the source and metabolism of cambial PGRs is scanty and inconclusive for both conifers and hardwoods.

Most cambial PGR research has focused on IAA. Much evidence indicates that this PGR is transported primarily in the cambial region at a rate of about 1 cm h^{-1}, and that the transport is basipetally polar. GC-MS measurements have established that endogenous IAA levels in the cambial region of Pinaceae are highest during earlywood development, and that cambial IAA levels may be considerably lower in hardwoods than in conifers. IAA appears to be involved in the control of cambial growth in conifers and hardwoods in at least three specific ways, viz. maintenance of the elongated form of fusiform cambial cells, promotion of radial expansion in primary walls of cambial derivatives, and regulation of reaction wood formation. In addition, it is well established that exogenous IAA promotes vessel development in hardwoods. In both conifers and hardwoods, exogenous IAA stimulates cambial growth in 1-year-old shoots treated late in the dormant period or after the start of the cambial growing period. However, exogenous IAA has little effect on cambia that are older or are in what is hypothesized to be the resting stage of dormancy. Thus it is uncertain whether IAA is directly involved in the control of cambial growth, or acts indirectly through a process such as hormone-directed transport.

It is not yet clear if gibberellins play a role in the control of cambial growth in conifers. However, in hardwoods, there is evidence that they inhibit vessel development and act synergistically with IAA in promoting cambial activity and fiber elongation. In both conifers and hardwoods, foliar sprays of gibberellins increase the accumulation of biomass above-ground, particularly in the main axis, while decreasing it in the roots.

There are as yet no definite conclusions to be drawn concerning the involvement of ABA, cytokinins, and ethylene in the regulation of cambial growth in conifers or hardwoods. In conifers, ABA may antagonize the promotory effect of IAA on cambial cell division and tracheid radial expansion under conditions of water stress, but high endogenous ABA levels do not appear to be associated with the formation of latewood or the onset of cambial dormancy. Some evidence suggests that exogenous cytokinins enhance the promotory effect of IAA on cambial growth, particularly ray formation, in both hardwoods and conifers. However, exogenous cytokinins, by themselves, appear to be ineffective. In hardwoods, ethylene-generating compounds satisfy the chilling requirement of the dormant cambium and promote the formation of wood having an apparently greater content of lignin and extractives. Ethylene-generators also affect wood development in conifers and accelerate cambial growth at the application site in both hardwoods and conifers.

1. INTRODUCTION

This review assesses the involvement of auxins, gibberellins (GAs), cytokinins (CKs), ethylene, and abscisic acid (ABA) in controlling the activity of the vascular cambium and the development of xylem and phloem in stems of tree species native to northern temperate regions. Cambial growth, in this review, refers to an increase not only in the cambium itself as the stem enlarges in circumference, but also in secondary vascular development as the stem enlarges in diameter. The former is accomplished by anticlinal division, while the latter involves periclinal division followed by differentiation of derivative cells into elements of xylem and phloem. Division occurs in elongated fusiform cells and nearly isodiametric ray cells, which comprise the cambial zone (172, 219). Cambial growth is seasonal, the cambium passing through an annual cycle of activity and dormancy. The extent of cambial growth and nature of vascular development vary with genotype, position in the tree, and environmental conditions (12, 31, 41, 52, 58, 85, 94, 192).

The hypothesis that cambial growth is under the control of plant growth regulators (PGRs) developed from the early finding that cambial reactivation in diffuse-porous hardwood species spread basipetally from expanding buds, and that debudding, defoliating, and girdling stopped cambial growth below the site of treatment. This research, and related work that investigated cambial reactivation and established indol-3-ylacetic acid (IAA) as a cambial stimulator in hardwoods and conifers, have been extensively reviewed (20, 95, 154, 162, 175, 189, 209, 211), and will be considered only briefly here. More recent reviews pertaining to PGR control of cambial growth in trees were prepared by Creber and Chaloner (31) and Savidge (172).

2. COMMENTS ON METHODOLOGY

Experiments designed to demonstrate that a particular PGR has a role in controlling cambial growth have generally followed two approaches, both of which have advantages and limitations. The first approach involves the application of authentic PGR(s) to variously debudded, defoliated, and derooted stems using a carrier such as lanolin or water, or as part of a chemically defined nutrient medium. The treated stems are grown under favorable environmental conditions and after several weeks the cambial response is assessed, typically in terms of cell type and xylem cell number, radial diameter, and secondary wall thickness. This procedure has the advantages of being relatively simple to perform and of enabling the investigator to apply a specific PGR directly to the cambium. However, the other tissues that make up the stem also are exposed to the PGR, and therefore it cannot be assumed that the cambial growth response necessarily is caused by direct action of the PGR. Alternatively, the response may be induced by biosynthetic products arising from altered metabolic processes in noncambial cells. Moreover, the cambium's response to exogenous PGR reflects the ability of cells in the cambial zone and other tissues to absorb, translocate, and metabolize the PGR. Unfortunately, these processes are affected by many interacting intrinsic and extrinsic factors, the most obvious being genotype, stem nutritional status, phenology, concentration and volume of application, carrier formulation, frequency and site of application, distance between the sites of application and response measurement, duration between initial application and the response measurement, and environmental conditions at and following the time of application. Consequently, it is not surprising that the literature relating to exogenous PGR effects on cambial growth contains many conflicting results.

The second experimental approach used to implicate a PGR in cambial growth control involves the measurement of endogenous levels in the cambial region (i.e., the cambial zone and its differentiating derivatives). This approach serves to confirm that the PGR occurs naturally and provides direct evidence for the existence of any correlative relationship between PGR levels and particular aspects of cambial growth. Unfortunately, the sophisticated analytical chemistry techniques (17, 73, 153, 234) that are required to measure endogenous PGR levels accurately and precisely drastically limit the number of laboratories able to do this research, and those having these facilities rarely investigate tree species. Thus, it is perhaps not unexpected that the literature pertaining to PGR measurements in plants contains few references to levels in the cambial region of trees. Most of the relevant papers report results obtained by bioassaying crude extracts, a relatively simple and inexpensive procedure. However, it is now appreciated that bioassay data have limited reliability both quantitatively and qualitatively, not only because bioassays are, at best, only approximately quantitative, but also because an unpurified extract can contain more than one active substance (10, 73). Reliable measurement necessitates the use of rigorous purification techniques, a specific physicochemical (73, 153) or immunological (214) assay, and an internal standard to account for the loss of endogenous PGR that inevitably occurs during purification and measurement. In addition, it is essential that the tissue under study for PGR levels be free of cells from adjoining tissues (178). In this review, emphasis is placed on studies involving quantitative PGR measurements that meet or closely approximate these criteria. Nevertheless, it must be emphasized that even when all of the above criteria are met, the significance of measuring cambial PGR levels remains unclear because (1) even cambial zone tissue contains cells of more than one type, and ideally the PGR level should be measured in a specific compartment of one cell type; (2) seasonal trends in cambial PGR levels typically are deduced from samples obtained from one or several trees that are sampled only once, and between-tree variation has been demonstrated for cambial growth (85) and PGR levels (170, 193); (3) the level of a PGR gives no indication of the rates of its synthesis, transport, catabolism, and storage, which may be more important than pool-size per se; and (4) changes in cambial sensitivity to the PGR may occur (57, 110, 202).

3. IDENTIFICATION, SOURCE, METABOLISM, AND TRANSPORT OF ENDOGENOUS PGRS IN THE CAMBIAL REGION

Although bioassays supply considerable evidence for the existence of promoting and inhibiting substances in the cambial region, there has been no full molecular characterization of a PGR in that region in any hardwood species, and only a few in conifers (Table 1). At present, only for some species of the Pinaceae can it be concluded that representatives of the auxin (IAA), inhibitor (ABA), cytokinin (Z, ZR, DCA), and gibberellin (GA_1, GA_4, GA_7, GA_9, GA_{20}) classes of PGR occur naturally in the cambial

TABLE 1. Identification by mass spectrometry of endogenous PGRs in vegetative parts of tree species

Compound	Species	Part	Reference
IAA	*Pinus radiata*	Shoots	235
	Pseudotsuga menziesii	Shoots	25, 43
		Shoots, xylem sap	34
	Picea sitchensis	Cambial region	112
	Larix decidua	Cambial region	176
	Pinus contorta	Cambial region	176
	Salix pentandra	Shoots	80
	Pinus sylvestris	Seedlings	168
		Cambial region	191
	Abies balsamea	Cambial region	193
ABA	*Acer pseudoplatanus*	Leaves	30
	Abies balsamea	Buds	113
	Pinus radiata	Stems	79
	Pseudotsuga menziesii	Leaves	13, 119
		Shoots	212
	Juglans spp.	Leaves	184
	Picea sitchensis	Cambial region	112
	Picea abies	Shoots	4
	Pinus sylvestris	Shoots	4, 72
	Juglans regia	Xylem sap	36
	Salix pentandra	Shoots	81
Z	*Acer pseudoplatanus*	Xylem sap	152
	Abies balsamea	Cambial region	111
	Castanea spp.	Shoots	233
	Gymnocladus dioica	Xylem sap	66
ZR	*Acer pseudoplatanus*	Xylem sap	74, 152
	Picea sitchensis	Leaves	116
	Abies balsamea	Cambial region	111
	Castanea spp.	Shoots	233
	Pinus radiata	Buds	196
	Gymnocladus dioica	Xylem sap	66
	Pinus sylvestris	Cambial region	191
DHZ	*Acer pseudoplatanus*	Xylem sap	152
DHZR	*Gymnocladus dioica*	Xylem sap	66
I^6Ade	*Castanea* spp.	Shoots	233
	Gymnocladus dioica	Xylem sap	66
I^6Ado	*Castanea* spp.	Shoots	233
	Gymnocladus dioica	Xylem sap	66
N^6-(2-hydroxybenzyl)adenosine			
	Populus x *robusta*	Leaves	75
DCA	*Acer saccharum*	Xylem sap	98
	Pinus contorta	Cambial region	172A

TABLE 1. (Continued)

Compound	Species	Part	Reference
GA_1	Salix pentandra	Shoots	37
	Pinus sylvestris	Cambial region	191
GA_4	Pinus sylvestris	Cambial region	191
GA_7	Pinus sylvestris	Cambial region	191
GA_9	Picea abies	Shoots	135
	Pinus sylvestris	Cambial region	191
iso-GA_9	Picea sitchensis	Leaves	114
GA_{19}	Juglans regia	Xylem sap	36
	Salix pentandra	Shoots	37
GA_{20}	Salix pentandra	Shoots	37
	Pinus sylvestris	Cambial region	191
GA_{29}	Salix pentandra	Shoots	37

Abbreviations: ABA = abscisic acid; DCA = dihydroconiferyl alcohol; DHZ = dihydrozeatin; DHZR = dihydrozeatin riboside; GA = gibberellin; IAA = indol-3-ylacetic acid; I^6Ade = isopentenyladenine; I^6Ado = isopentenyladenosine; Z = zeatin; ZR = zeatin riboside.

region. Evidence for the occurrence of IAA in the cambial region of hardwoods is provided by Savidge and Wareing (176) who used the combined gas chromatography-mass spectrometry (GC-MS) technique of single ion current monitoring (SICM) to quantify IAA in that region in Acer pseudoplatanus L., Populus x canadensis 'Robusta', and Quercus robur L. Gas chromatographic measurements indicate that ethylene is produced by stems of both hardwoods and conifers (14, 24, 69, 99, 125, 126, 143, 159, 199), but there is no conclusive evidence that ethylene occurs in the cambial region per se. PGR distribution within the cambial region has been convincingly determined in only one investigation. Savidge et al. (178) observed that the bulk of the IAA that was present in the cambial region of Pinus contorta Dougl. during the growing season occurred in the zone where cambial derivatives differentiate into xylem elements.

Little information is available concerning where cambial PGRs are synthesized and how they are metabolized. It has long been speculated that the cambial region itself produces PGRs (185, 189, 203), but it has yet to be demonstrated with enzyme preparations that cambial region cells can in fact convert precursors into PGRs or PGRs into conjugates or catabolites. Only one PGR intermediate, that of ethylene, has been identified conclusively in the cambial region (179). Pharis et al. (139) observed that cambial scrapings contained several unidentified GA-like compounds 14 days after ^{3}H-GA_4 was injected into Pinus radiata D. Don stems. Riov and Gottlieb (157) found that incubating Pinus pinea L. shoot bark and shoot wood for 24 h with 2-^{14}C-IAA resulted in the formation of indol-3-acetylaspartic acid (IAAsp) and IAA-glucoside. Furthermore, 2-^{14}C-IAA was converted into indol-3-ylmethanol (IM) in short-term experiments with protoplasts (166) and crude enzyme preparations (194) from Pinus sylvestris L.

shoots. Additional unequivocal identifications that may pertain to cambial region PGR metabolism are Z-glucoside in Castanea spp. shoots (233), and IAAsp (3), ABA-glucoside (72), and IM (195) in Pinus sylvestris shoots.

Apical meristems and leaves are likely sources of PGRs to the cambial sink. This supposition is supported by the unequivocal identification of IAA, ABA, CKs, and GAs in buds, expanding shoots, xylem sap, and leaves of many tree species (Table 1). In addition, there are related, conclusive characterizations of GA_9-glucoside in Picea sitchensis (Bong.) Carr. leaves (115), dihydrophaseic acid in Juglans regia L. xylem sap (36), indole-3-ethanol (164) and indole-3-carboxylic acid (169) in Pinus sylvestris leaves, and ZR-glucoside in Pinus radiata buds (196). Treatments that removed the influence of buds, shoots, and leaves (debudding, decapitating, defoliating, and ring girdling) decreased the level of IAA (35, 110, 176, 220) and ABA (110) in the cambial region below the treatment site. The contents of IAA and ABA in current-year Pinus contorta leaves decreased as the leaves matured, attaining the levels found in 1-year-old leaves by late summer (177). High IAA and ABA levels were also found in current-year and 1-year-old leaves of Picea sitchensis (110). In addition, Little and Wareing (110) observed that water stress decreased the IAA level and increased the ABA level in current-year leaves more than it did in 1-year-old leaves. In a comprehensive study of pool size, metabolism, and transport of IAA in 8-week-old Pinus sylvestris seedlings, the IAA level was higher in the leaves than in the shoot-axis, hypocotyl, cotyledons, or roots, and the leaves exported more radioactivity than any of the other four parts of the seedling when ^{14}C-IAA was applied separately to each part (167). These findings with tree species support the notion that apices and leaves, particularly young leaves, are important sites of IAA and ABA synthesis, as they are thought to be in other species (101, 181).

The root apex is another possible source of cambial PGRs, and experiments conducted with nonwoody species suggest that roots not only are the major sites of CK synthesis, but also are important producers, together with young leaves, of GAs (101, 181) and the ethylene precursor 1-aminocyclopropane-1-carboxylic acid (ACC;232). Mycorrhiza-forming fungi, Rhizobium bacteria, and certain phytopathogenic microorganisms also produce PGRs (10, 49, 71, 128), but whether they and other microorganisms such as mycoplasma and xylem-limited bacteria supply PGRs to the cambial region remains to be determined.

IAA, ABA, GAs, and CKs have been detected in the phloem and xylem saps of many tree species (Table 1; 82, 215), and radial transport across the cambial region between the xylem and phloem has been observed for ^{14}C-IAA in Salix viminalis L. (56, 100), Salix fragilis L. (55), and Acer pseudoplatanus (243), ^{14}C-kinetin (K) in Salix viminalis (16, 100), and ^{14}C-GA_3 in Salix viminalis (16) and Salix fragilis (55). These observations suggest that PGRs circulate in the vascular system, but with the notable exception of IAA, little is known about PGR transport within the cambial region of trees.

Much evidence indicates that IAA moves primarily in the cambial region and that its transport is basipetally polar. In experiments with stems of Malus sp. (5), Abies balsamea (L.) Mill. (104, 105), Populus deltoides Bartr. ex Marsh. (132), Pinus densiflora Sieb. and Zucc. (134), Pinus taeda L. (236), Pinus jeffreyi Grev. and Balf. (236), and Pinus ponderosa Laws. (236), ^{14}C-IAA was readily transported and movement in the basipetal direction exceeded that in the acropetal direction. Microradiographs of cross-sections of Fagus sylvatica L. shoots supplied apically with ^{3}H-IAA indicated that IAA transport was localized in the cambial zone, differentiating derivatives, and phloem parenchyma during cambial activity, but occurred in the cambial zone and phloem parenchyma during cambial dormancy (89, 90). Similarly, Nix and Wodzicki (130) found that the cambial zone and differentiating derivatives, particularly on the xylem side, contained most of the radioactivity recovered 2 days after ^{14}C-IAA was applied apically to growing Pinus echinata Mill. stems. Experiments with Pinus taeda (22), Picea sitchensis (112), Malus sp. (136), Pyrus sp. (136), and Pinus sylvestris (240) demonstrated that endogenous IAA was transported out of stem segments into agar, and that efflux from the basal end was greater than that from the apical end. In contrast, the efflux of endogenous ABA from the basal end of Picea sitchensis stem segments was similar to that from the apical end (112), which suggests that ABA movement is nonpolar. Apically applied IAA, compared with basal application, stimulated more cambial growth along the length of Abies balsamea (104), Pinus sylvestris (239), and Quercus robur (242) shoots. Similarly, laterally applied IAA stimulated more xylem production below the application point than above this point in Pinus halepensis Mill. (51), Salix sp. (53), and Abies balsamea (Fig. 1). Finally, girdling was shown to interrupt the transport of ^{14}C-IAA in Abies balsamea (105), of endogenous IAA in Prunus persica (L.) Batsch (35), Picea sitchensis (110), and Pinus contorta (176), and of endogenous ABA in Picea sitchensis (110).

Experiments with Malus sp. (62), Fraxinus americana L. (70), Fagus sylvatica (89), Abies balsamea (105), Pinus densiflora (134), and Acer pseudoplatanus (243) indicated that IAA is transported at a rate of about 1 cm h^{-1} when the cambium is active. The actual rate depends on the prevailing environmental conditions, as it was observed that IAA transport increased with increasing temperature up to a maximum of about 25°C in Malus sp. (62) and Fagus sylvatica (89), and that water stress decreased the rate of ^{14}C-IAA movement in Abies balsamea (104). IAA movement also depends on the physiological condition of the cambium, a decline in transport after the cambium enters dormancy having been documented in Fraxinus americana (70), Fagus sylvatica (89), Abies balsamea (105), Populus deltoides (132), Malus sp. (136), Pyrus sp. (136), and Acer pseudoplatanus (243). However, the decline occurs gradually, because it was found in Fagus sylvatica (89), Abies balsamea (105), and Pinus densiflora (134) that the transport of labeled IAA for several weeks after the cessation of periclinal division and xylem development was comparable to that during cambial growth. This finding indicates that maximal IAA movement in the cambial region does not require the presence of differentiating derivatives (89, 90, 105).

Zajączkowski and Wodzicki (240), using a bioassay, found that the quantity of IAA diffusing basipetally from consecutive short segments of Pinus sylvestris stems oscillated in a wave-like pattern. Subsequent experiments established that similar patterns were also present in Larix decidua Mill., Acer pseudoplatanus, Quercus robur, and Picea abies (L.) Karst., that the amplitude of the wave was altered by applying IAA, ABA, or 2,3,5-triiodobenzoic acid (TIBA, an inhibitor of IAA transport), and that the growth promotor active in the bioassay was actually IAA (222, 224, 225, 242). This and other evidence (241) led these investigators to hypothesize that the cambial region functions as a supracellular oscillatory system generating IAA waves, and that these waves specify morphogenic positional information to differentiating derivatives. However, the finding that the catalytic capacity of various leaf enzymes oscillated spontaneously under constant conditions (152A) suggests alternatively that the IAA waves may be generated at extra-cambial sites of synthesis.

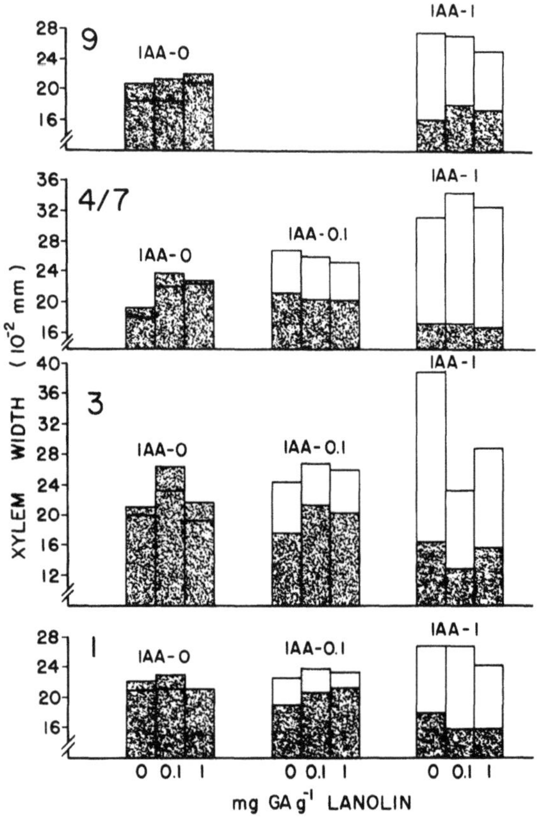

FIGURE 1. Width of new xylem measured 1 cm above (stippled bars) and below (open bars) the point of lateral application of GA_1, GA_3, $GA_{4/7}$, or GA_9 to the terminal shoot of 3-year-old Abies balsamea seedlings. Each GA was applied with 0, 0.1, or 1.0 mg IAA g^{-1} lanolin. The seedlings were dormant at the start of the experiment on April 11, and were grown for 6 weeks in controlled environmental conditions fávorable for growth. Analysis of variance indicated that IAA increased and decreased tracheid production below and above the application point, respectively, whereas GA had no effect except for the GA_3+1 mg IAA treatments, which were inhibitory. It was observed in additional experiments that neither apically nor basally applied GA_3 affected tracheid production in cuttings supplied apically with 0, 0.1, or 1 mg IAA g^{-1} lanolin (CHA Little, unpublished).

4. PHYSIOLOGICAL SIGNIFICANCE OF PGRS IN THE CAMBIAL REGION

4.1. Conifers

4.1.1. Auxin. Experiments with 1-year-old foliated shoots indicate that IAA is involved in the regulation of cambial activity (i.e., mitotic activity) and xylem development. The main evidence for this statement is the combined finding in many species that debudding, defoliating, or girdling decreased the cambial IAA level and halted cambial reactivation in chilled dormant shoots, and cambial growth in active shoots, immediately below the treatment point, whereas restoring the supply with exogenous IAA prevented the decline in cambial IAA level and enabled cambial growth to proceed (6, 42, 68, 92, 96, 106, 110, 131, 151, 174, 176, 186, 220, 223, 226, 238, B Sundberg and CHA Little, unpublished). The concentration of applied IAA, up to an optimal level, was positively correlated with tracheid production in Abies balsamea (106), Picea sitchensis (110), Pinus radiata (186), and Pinus sylvestris (238), suggesting that the rate of cambial derivative production is influenced by the level of endogenous IAA. This notion is supported in general by the finding of a higher cambial IAA level during xylem development than when the cambium was dormant in Picea sitchensis (110), Pinus sylvestris (165), Pinus contorta (177), and Abies balsamea (193). However, Savidge et al. (178) found in Pinus contorta that the change in endogenous IAA level between activity and dormancy occurred mainly in the zone of developing xylem and that there was no evidence of a correlation between the number of cells per radial file in the cambial zone and the content of endogenous IAA in the cambial zone.

It has been demonstrated with many species that exogenous IAA promotes the radial enlargement of primary-walled cambial derivatives on the centripetal side of the cambial zone, i.e., "tracheid radial enlargement," and, less decidedly, the thickening of the secondary wall in these cells, i.e., "tracheid secondary wall thickening" (42, 59, 68, 78, 92, 93, 131, 151, 171, 175, 186), which suggests that these developments are affected by the level of endogenous IAA. Involvement of endogenous IAA in the control of tracheid radial enlargement, but not in the regulation of tracheid secondary wall thickening, is indicated by the finding of a temporal correlation between declining cambial IAA levels and the annual transition from large-diameter thin-walled earlywood tracheids to narrow-diameter thick-walled latewood tracheids in Picea sitchensis (110), Pinus sylvestris (165), Pinus contorta (177, 178), and Abies balsamea (193). Similarly, Little and Wareing (110) observed in a droughting experiment with Picea sitchensis seedlings that water stress decreased the cambial IAA content and induced the formation of narrow-diameter tracheids. Moreover, tracheid radial enlargement in Pinus resinosa Ait. was inhibited by TIBA (95). All of the preceding evidence contradicts the recently proposed hypothesis (1, 2) that xylem cell size is negatively correlated with IAA level at the time of differentiation. In support of this concept, Saks and Aloni (163) reported that tracheid radial diameter increased basipetally along the hypocotyl of Pinus pinea seedlings, but they made no IAA measurements in an attempt to confirm their speculation of an associated decreasing gradient in IAA content.

Several findings demonstrate that the cambial response to IAA varies with cambial age and degree of foliation, which indicates that factors in addition to IAA are involved in the control of cambial growth. In experiments with defoliated Pinus contorta stem segments of different ages, exogenous IAA promoted cambial cell division and tracheid differentiation along the entire length of the treated shoot only in 1-year-old material (171, 174). In 2-year-old cuttings, apically applied IAA stimulated cambial activity and tracheid differentiation near the application point and at the cutting base, but not in the intervening region, although the IAA must have been transported through it. IAA, applied alone or together with other PGRs, did not induce cambial growth anywhere in 3-year-old and older stem segments or cultured cambial explants. Nevertheless, it was observed in these materials, as well as in younger defoliated stem segments, that IAA prevented the fusiform cells of the cambial zone from dividing transversely and differentiating into axial parenchyma, which indicates that IAA is required to maintain the morphological state of these cells (171, 175). The promoting effect of exogenous IAA on tracheid production also decreased with increasing cambial age in Abies balsamea (156) and Pinus sylvestris (238). Exogenous IAA promoted periclinal cell division and tracheid radial enlargement over the length of 1-year-old cuttings of *Pinus contorta* (174) and other conifer species (171) kept in the light and defoliated except for a single foliar organ near the midpoint; however, complete tracheid differentiation occurred only in the immediate vicinity of the junction with the leaf vascular system. The initiation of cambial cell division and tracheid radial enlargement during springtime cambial reactivation in field-grown *Pinus contorta* trees was spatially and temporally separated from that of bordered-pit development and secondary wall formation (175, 177). In addition, high cambial IAA levels were found below the live crown in May while the cambium was dividing rapidly and its numerous derivatives were in a state of primary-wall enlargement, yet secondary wall development was absent (177). All of these observations considered together indicate that other biochemical factors besides IAA are active in the control of cambial growth, and, more specifically, that mature leaves produce a tracheid differentiation factor, which remains to be characterized (174).

The notion that IAA is not the only factor regulating cambial growth is further supported by observations indicating that the cambial response to IAA varies with season, *i.e.*, phenology. IAA application to *Pinus sylvestris* shoots late in the cambial growing period stimulated less tracheid production than did earlier treatment under both naturally changing and artificially constant environmental conditions (223, 226, 238, 239). Exogenous IAA did not prevent the cessation of cambial activity under natural environmental conditions in *Picea abies* (42) and *Picea sitchensis* (110) shoots. Similarly, exogenous IAA did not stop the onset of cambial dormancy induced in active *Picea sitchensis* seedlings by artificial short-day photoperiod despite temperatures favorable for growth (110). In experiments with dormant, debudded shoots of *Pseudotsuga menziesii* (Mirb.) Franco (96), *Pinus densiflora* (131), and *Abies balsamea* (105, 106, 155, 156, 193) grown under controlled environmental conditions favorable for growth, apically applied IAA induced less tracheid production at the beginning of the dormant period than at the end. The transport of ^{14}C-IAA early in the dormant period in *Abies balsamea* (105) and *Pinus densiflora* (134) was comparable to that during the cambial growing period, indicating that the cambium's relative lack of response to exogenous IAA at the beginning of dormancy cannot be attributed to insufficient IAA transport.

The above observations, together with the findings that IAA was present in large amounts at the beginning of the cambial dormancy period (110, 165, 176, 177, 193), and that the cambial response to exogenous IAA under favorable growing conditions in dormant *Abies balsamea* shoots increased with increasing duration of pre-exposure to natural and artificial chilling (106, 155), prompted two hypotheses pertaining to the regulation of the annual cycle of cambial activity and dormancy. The first (106) proposes that at the beginning of the dormant period, the cambium, like buds, is in a state of rest, which is maintained by internal factors and broken by chilling (91, 138, 161). However, no evidence for a cambial chilling requirement was obtained in experiments with *Pseudotsuga menziesii* (97, 228). Moreover, cambial growth was observed during midwinter in *Pinus taeda* and *Pinus elliottii* Engelm. seedlings in the southeastern United States (21, 129) and in *Pinus radiata* trees in New Zealand (8). However, these observations were made with species native to climates having winter temperatures that are relatively mild compared with those experienced by *Abies balsamea* growing in the continental climate of eastern Canada. The evidence suggests that there is genotypic variation in the occurrence of cambial rest (8), the chilling requirement being manifested only in cold climates where the development of cambial frost hardiness is a desirable adaptive trait (155). The second, related hypothesis (110, 193) proposes that the cambium's annual transition from activity to rest, where it occurs, is due not to a deficiency in IAA, but to the development of an unresponsiveness to IAA, and that the gradation from rest to quiescence, during which state dormancy is imposed only by low temperatures, involves regaining the ability to respond to IAA. Although the cause of the resting cambium's inability to respond to IAA has yet to be determined (155, 156), it is evident from experiments involving the localized application of heat to dormant stems located in the cold that cambial quiescence is in fact maintained by low temperatures (175, 229). Temperature may also affect the pattern of cambial reactivation, which has been reported both to progress basipetally from bud bases in association with bud burst, and to begin a considerable distance below the bud bases independently of bud burst (see 175).

In many species, the application of a relatively high concentration of IAA to vertically oriented stems induced the formation of compression wood (15, 59, 68, 78, 93, 124, 131, 151, 171, 186, 190, 206, 216), which typically occurs on the lower sides of branches and inclined stems (218). Furthermore, applying morphactin IT 3456, an inhibitor of IAA transport, to a vertically oriented or inclined stem promoted compression wood formation above the application point, and, in inclined stems, inhibited such formation below (142, 144, 187, 230, 231). These observations suggest that compression wood formation is associated with high levels of endogenous IAA, but the validity of this suggestion needs to be tested using a reliable procedure for measuring IAA.

It is well established that altering the normal axial direction of IAA transport by partial girdling, leaving a narrow diagonal bridge of intact tissues, causes the fusiform cambial cells in the bridge to reorient until their long axes parallel the bridge angle (63, 64, 65, 173, 237), which suggests that the angle of IAA flow influences the orientation of the fusiform cambial cells (63, 65, 237). This hypothesis is supported by the observation that the reorientation of fusiform cambial cells near the upper edge of a bridge in helically girdled stems of _Pinus radiata_ (63) and _Abies balsamea_ (237) was inhibited by application of indol-3-ylbutyric acid (IBA) and IAA, respectively, to the girdle. In addition, Zagórska-Marek and Little (237) found that ^{14}C-IAA transport was inhibited in a helical bridge in current-year _Abies balsamea_ shoots. The extent of this inhibition was markedly greater 1 day after the girdling than at the end of a subsequent 11-week growing period, during which interval considerable fusiform cambial cell reorientation occurred near the bottom edge of the bridge. However, it is evident that abundant periclinal division and tracheid differentiation can occur in the diagonal bridge prior to any reorientation, although diagonal IAA transport must be occurring (173). Thus the control of fusiform cambial cell orientation involves factors in addition to IAA, one of which likely is physical force (19, 67, 102).

It is well established that auxin promotes the proliferation of callus in stem explants cultured _in vitro_ (23). Induction of tracheid-like cells and of nodules containing apparent xylem, phloem, and cambium have been observed in some callus subcultures (_e.g._, 9, 103). Sclereids rather than tracheids differentiated in callus cultures initiated from cambia of _Pinus contorta_ (171).

The production of vertical resin ducts in _Pinus halepensis_ stems was stimulated by laterally applied 1-napthaleneacetic acid (NAA), 2,4-dichlorophenoxyacetic acid (2,4-D) or IAA (51).

In summary, the current evidence indicates that a continuous supply of basipetally transported IAA is required to maintain fusiform cambial cell structure and to enable cambial growth to proceed, but IAA by itself is insufficient to promote mitosis, cytokinesis, and tracheid secondary wall differentiation, and to control fusiform cambial cell orientation. Evidence for a regulatory role of IAA level is consistent only for tracheid radial enlargement, supporting the many observations implicating IAA in the control of primary-wall extension in nonforest dicots (29, 50). The level of IAA may also control the formation of compression wood. The cambial response to exogenous IAA varies with the season of application and decreases basipetally from the application point and with increasing cambial age. The evidence can be interpreted as indicating that IAA indirectly controls cambial growth by directing the transport of substances that affect cambial growth more directly, the levels of which vary in time and space. These substances presumably include nonspecific assimilates (137, 226) and presumptive substances affecting specific aspects of cambial growth, e.g. tracheid differention (172, 174). The availability of other PGRs may also be involved. It is also possible that IAA regulates cambial growth through changes in the pattern of its transport (241) or in the sensitivity of the cambial zone and differentiating derivative cells to its presence (42, 57, 155, 202), but lack of evidence precludes further discussion.

4.1.2. Abscisic acid. Experiments with exogenous ABA indicate that cambial growth is inhibited by this PGR. ABA reduced the frequency of periclinal division and the extent of tracheid radial enlargement when basally supplied to *Picea glauca* (Moench) Voss cuttings (107) and injected into *Pinus radiata* seedlings (78, 139). Similarly, basal application of ABA decreased tracheid production in *Abies balsamea* cuttings (104, 108) and tracheid differentiation at the trace-cambium junction of *Pinus contorta* cuttings bearing a single needle fascicle (171). That ABA inhibits tracheid production by altering IAA transport is suggested by the findings that basally supplied ABA decreased the movement of a ^{14}C-IAA pulse in *Abies balsamea* cuttings (104, 105), and that the apical application of ABA reduced the amplitude of the IAA wave detected by basipetal efflux from successive stem segments in *Pinus sylvestris* (224).

The inhibitory effects of exogenous ABA on cambial growth support the hypothesis (110, 221) that the annual changeover from earlywood to latewood formation and eventual cessation of cambial mitotic activity are due to an accumulation of endogenous ABA. Webber et al. (212) provided some evidence for this hypothesis, finding in *Pseudotsuga menziesii* shoots that the ABA level, measured using gas chromatography and an internal standard, was higher during the midwinter dormant period than in July, when the cambium presumably was still active. However, results obtained in more definitive experiments clearly indicated that no specific aspect of cambial growth in stems of *Picea sitchensis* (110) or *Pinus contorta* (177) was associated with a particular change in cambial ABA level, with the possible exception of the production of narrow-diameter tracheids during false-ring formation. Working with field-grown *Picea sitchensis* trees, Little and Wareing (110) observed that two severe natural droughts coincided with a large increase in the cambial content of ABA, the first of which occurred during the cambial growing period while narrow-diameter tracheids were being produced temporarily. These investigators confirmed in an artificial droughting experiment with potted seedlings that high water stress markedly increased the cambial ABA level, inhibited tracheid production, and caused the formation of relatively narrow-diameter tracheids. The same stress, however, simultaneously decreased the cambial IAA level, as noted previously.

4.1.3. Gibberellins. Experiments with several species have established that exogenous GA_3 and $GA_{4/7}$ alter the normal distribution of assimilates, with cambial and longitudinal growth in the main shoot being promoted at the expense of growth in the branches and roots (11, 109, 162, 213). However, evidence implicating GA in the control of cambial growth *per se* is both meager and inconsistent, hence no conclusions can be drawn. Exogenous GA_3 enhanced the promoting effect of exogenous IAA on tracheid production in some studies (42, 68, 131, 211), but not in others (15, 42, 96, 106, 145, 223, 238, Fig. 1). All the investigators just cited reported that GA_3 by itself did not affect tracheid production or differentiation. In the only known investigation of the effect of a PGR on phloem development in a conifer, the application of GA_3 through a micropipette to the cambial zone of dormant *Pinus strobus* L. explants in sterile culture induced radial expansion and secondary wall development in immature sieve cells (39). In experiments with other GAs, lateral applications of GA_1, $GA_{4/7}$, or GA_9 applied alone or in combination with IAA had no effect on tracheid production in *Abies balsamea* (Fig. 1). However, injection of GA_4 and GA_4 + ABA increased tracheid radial enlargement in *Pinus radiata* (139).

4.1.4. Ethylene. The little evidence available suggests that ethylene is involved in the control of cambial growth, particularly in the differentiation of compression wood tracheids. The ethylene precursor ACC was found only on the lower side of *Pinus contorta* branches in association with compression wood formation (179). The application of ethrel (2-chloroethylphosphonic acid), an ethylene-generating compound, stimulated radial growth in vertical stems of *Pinus radiata* (7), *Picea strobus* (24), and *Pinus taeda* (199, 200), apparently by increasing the production of both xylem and phloem tissue (7, 199). Other observed effects included a decrease in tracheid wall thickness (7, 200) and a stimulation of the production of rays (7) and vertical resin ducts (199, 200). Ethylene evolution increased in *Cupressus arizonica* Greene branches released from apical control (14) and in *Pinus strobus* branches tied into loops (24); in both cases an increase in compression wood formation would also be expected. Moreover, the evolution of ethylene was greater from the lower half of the branch than from the upper half (14). Telewski and Jaffe (199) observed that ethylene production in *Pinus taeda* and *Abies fraser* (Pursh) Poir. shoots was enhanced by wind and by mechanically-induced flexing, which they found also increased tracheid production and decreased tracheid length (197, 198).

4.1.5. Cytokinins. This class of PGRs has been investigated the least in connection with cambial growth control, and more research is required before the extent of their involvement can be assessed. Ringing *Picea sitchensis* stems with N^6-benzyladenine (BA) alone or together with IAA stimulated the production of xylem, phloem, and ray tissue (145). Mixtures of K with IAA, NAA, or 2,4-D promoted ray formation in *Pinus halepensis* (51). Applying BA + IAA and K + IAA apically to *Pinus sylvestris* stems increased tracheid production, whereas BA and K alone did not (68). Similarly, Blum (15) observed that K + IAA, but not K by itself, stimulated compression wood formation when applied laterally to *Picea abies* stems. However, neither K nor K + IAA increased cambial growth when applied apically in other experiments with *Pinus sylvestris* shoots (223, 238) or to *Abies balsamea* cuttings (106). Tracheid production in *Abies balsamea* cuttings was not promoted by basally supplied Z, ZR, or BA in the presence of endogenous or exogenous IAA, and at high concentrations these

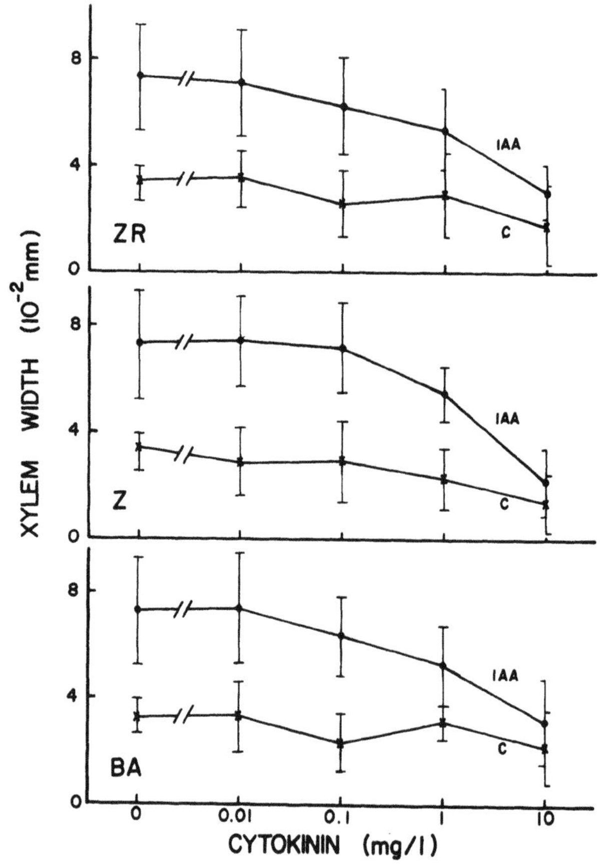

FIGURE 2. Effect of basal application of BA, Z, or ZR in aqueous solution on width of new xylem produced at the midpoint of budded control cuttings (C) and debudded cuttings supplied apically with 1 mg IAA g^{-1} lanolin (IAA). The cuttings were obtained October 31 from current-year shoots of dormant ca. 7-m-tall field-grown *Abies balsamea* trees and were cultured for 8 weeks in controlled environment conditions favorable for growth (CHA Little, unpublished).

cytokinins were inhibitory (Fig. 2). In the only investigation of seasonal changes in CK content in the cambial region, Kubowicz (86) detected fluctuations in bioassay activity in five fractions obtained from *Pinus sylvestris* stems.

4.2. Hardwoods

4.2.1. Auxin. Early attempts to promote cambial activity in young stem segments of hardwoods with exogenous IAA were not convincing, possibly because the cambium was in the rest state of dormancy. In cuttings of Alnus, Crataegus, Fraxinus, Gleditsia, Philadelphus, Populus, Salix, and Tilia spp., cambial reactivation occurred over a distance of only 3 cm or less basal to the site of IAA application (154). Gouwentak and Maas (61) deduced that the dormant cambium required chilling before IAA would reactivate it over the cutting's full length.

It is now well established for debudded 1-year-old cuttings collected near the end of the dormant period, after the presumptive rest-quiescence transition has occurred, that exogenous auxin applied to the apical end or tangential surface of both diffuse- and ring-porous hardwoods stimulates cambial activity and xylem development, notably vessel development, basal to the application site (6, 45, 46, 84, 120, 242). In addition to promoting vessel development per se, IAA stimulates the radial expansion of developing vessel elements, particularly in ring-porous species (45, 46, 133, 185, 188).

Very little auxin research has been conducted with older cambia. McQuilkin (117) applied IAA, IBA, and NAA to large trunks of Fagus, Acer, Nyssa, and Quercus spp. three times during the year (fall, winter, and spring), but recorded no promotory effects. In contrast, unspecified auxin promoted the development of vessels, but not of fibers in 4- to 10-cm-diameter trunks and branches of Robinia pseudoacacia (84). IAA stimulated cambial activity and xylem development in 3- to 5-year-old lateral shoots of 20-year-old Quercus robur (242). Brown and Cormack (18), working with 3-year-old cambium of Populus balsamifera L., found that although IAA promoted cambial activity immediately below the site of application and also above a ring located some distance basal to the IAA application site, it did not induce cambial growth in the intervening stem tissue.

There is additional, mostly indirect, evidence for a regulatory role of auxin in cambial growth. High IAA concentrations decreased fiber length in Populus robusta (45), and in the absence of IAA, very short xylem elements were formed in stem segments of Populus robusta (45) and Robinia pseudoacacia L. (84). NAA promoted cambial activity and an abnormal wood containing abundant parenchyma and short, thin-walled tracheids in 2-year-old Malus sp. seedlings (146). In Prosopis juliflora (Swartz) DC var. glandulosa (Torr.) Cockrell, cambial derivatives differentiated into axial parenchyma below the point of application of TIBA, whereas 2,4-D or 2,4,5-trichlorophenoxyacetic acid (2,4,5-T) at 2% (w/w in lanolin) caused formation of numerous narrow-diameter, thick-walled vessel elements at 1 mm below the treatment site (160). Morphactin IT 3456 greatly reduced vessel diameter and the lengths of vessel elements, tracheids, and fibers in Carpinus betulus L. and Syringa vulgaris L.; however, secondary wall thickening was increased (188). The same morphactin inhibited tracheid expansion and reduced lignification in Salix fragilis L. (141) and Morus alba L. (182). Sharma et al. (183) observed that morphactin IT 3456 also caused cambial ray cells of Morus alba to differentiate into sieve cells.

Observations on the effects of auxin at the subcellular level are few. NAA caused a complete disappearance of starch in both chilled and unchilled Malus sp. shoots (148). IAA promoted the incorporation of ^{14}C-glucose into cambial derivatives of Populus deltoides (217) and inhibited the formation of wound callose in sieve elements of Salix viminalis (201).

Several observations indicate that an auxin deficiency results in the differentiation of tension wood fibers (26, 123). In inclined stems of Aesculus hippocastanum L. (26), Acer rubrum L. (33), and Populus deltoides (124), the development of tension wood was suppressed by the application of auxin to the upper side of the stem. Application of TIBA caused differentiation of tension wood fibers basal to the application site in upright

Acer rubrum (32) and Ulmus americana (83) stems. IAA, NAA, or 2,4-D applied to the TIBA treatment site suppressed tension wood development and permitted tracheid differentiation in upright Acer rubrum seedlings (122). Phelps et al. (140) found that the application of morphactin IT 3456 to the convex side of artificially bent branches of Aesculus hippocastanum prevented tension wood formation, promoting instead the production of rays and the differentiation of shortened vessel elements and fibers. Application of IAA or 2,4-D to one side of upright Aesculus hippocastanum epicotyls induced formation of tension wood on the side opposite the application site (26), and upright Acer rubrum seedlings produced tension wood in response to low concentrations of NAA and 2,4-D (123). Thus, Casperson (26) and Morey and Cranshaw (123) suggested that tension wood development is fostered by a gradient in auxin concentration, rather than by merely a low level of auxin.

In diffuse-porous species, cambial reactivation and first earlywood formation typically have been reported (154, 207, 209) to begin at the bud bases at the time of budbreak, and to proceed basipetally along young shoots and branches at about 1 cm h^{-1}, which, as noted previously, is the established rate of basipetal IAA transport. However, Lachaud and Bonnemain (88) found that cambial growth occurred simultaneously throughout the bole of Fagus sylvatica. Cambial growth stops very soon after the cessation of extension growth and leaf development, and after defoliation. If a diffuse-porous hardwood is debudded in early spring, cambial growth does not occur until new buds grow out; however a notable exception was documented at bases of debudded Prunus armeniaca L. trees (47). Following the removal of a ring of bark and cambium, cambial activity and wood formation continue above the girdle. Cambial growth does not occur below the girdle, except for localized activity within a few centimeters of the wound (18).

Cambial growth in ring-porous hardwoods contrasts markedly with that in diffuse-porous species. Treatments such as debudding, ringing, or cutting the stem of even a large-diameter tree into short lengths do not prevent development of the first earlywood vessels (40, 207, 209, RA Savidge and PF Wareing, unpublished). Cambial reactivation and first earlywood development have often been reported to occur simultaneously over the entire stem and throughout the branches of even large trees. However, a basipetal progression in cambial reactivation was evident in Quercus sp. (88) and Ulmus americana L. (150). The abrupt cessation of wide-diameter, earlywood vessel production coincides with the cessation of shoot elongation and leaf development (40, 209, 210). However, wood formation continues for a considerable time thereafter, which suggests that extending shoots and developing leaves export IAA to the cambium, whereas mature leaves produce a cambial stimulus different from IAA (210).

Many bioassay measurements have indicated that endogenous auxin levels increase in the cambial region in association with budbreak and leaf expansion (38, 175). However, RA Savidge (unpublished), using GC-MS in the SICM mode, found that IAA levels in the cambial region of 1-year-old and older main-stem material from Acer pseudoplatanus, Populus x canadensis 'Robusta' and Quercus robur saplings did not increase during cambial reactivation and leaf expansion. The IAA levels were near 100 ng g^{-1} dry weight in these hardwoods, which contrast with the high IAA levels (1-20 µg g^{-1}) found in conifers using the same SICM technique (176, 177). Dann et al. (35) also detected similarly low IAA levels in the active cambial region of Prunus persica, using the indolo-alpha-pyrone fluorescence method of measurement.

Tissue culture investigations with both ring- and diffuse-porous hardwoods have shown that the callus generated by cambial explants can differentiate into xylem elements on media lacking auxin or other hormones (77; CL Brown, personal communication). Moreover, Jacquiot (76) found that several auxins, including IAA, were ineffective in promoting xylem development in cambial cultures of many species. Venverloo (204) described callus cultures of Populus nigra L. cv 'Italica' that gradually lost the capacity to differentiate tracheids on a medium containing 2,4-D, but that recovered this ability when subcultured on a medium containing IAA.

In summary, there is considerable evidence that auxin levels regulate the extent of radial enlargement of cambial derivatives, the number of vessels that develop, and the occurrence of tension wood formation in young shoots of both ring- and diffuse-porous hardwoods. Some evidence suggests that auxin, by its presence, maintains the elongated fusiform shape of cambial cells. It remains unclear if the occurrence of cambial cell division near the site of auxin application is due to direct action of auxin on cell-division processes. Endogenous IAA, if it exists at all in the cambial region of hardwoods, appears to occur at levels many times lower than those found in conifers.

4.2.2. Abscisic acid. We know of only one investigation concerning ABA and the hardwood cambium. Feucht et al. (54) provided gas chromatographic evidence for ABA in the cambium of three Prunus spp. In the presence of ABA, shoot segments of all three species produced cambial callus in vitro. This callus grew well in winter and summer but poorly in October.

4.2.3. Gibberellins. Wareing (208) demonstrated that the application of GA_3 to the apical end of debudded shoots of Acer pseudoplatanus, Populus nigra v. italica, and Fraxinus excelsior L. promoted cambial cell division and derivative expansion, but not normal vessel or fiber development. Application of GA_3 around the circumference of intact Fraxinus americana seedlings promoted the production of cambial derivatives, but inhibited lignification (as indicated by staining with phloroglucinol-hydrochloric acid) and vessel element differentiation (RA Savidge, unpublished). GA_3 also stimulated cambial activity below the application site in Acer rubrum (121). Zakrzewski (242) observed that GA_3 promoted cambial activity in 3- to 5-year-old cuttings of Quercus robur, and that whereas IAA was effective only when applied at the apical end, GA_3 promoted cambial activity when introduced at either end. In contrast, GA_3 did not stimulate cambial activity in Fagus sylvatica (87), Malus sp. (149), and Salix fragilis (158). Moreover, the gibberellin antagonists (2-chloroethyl)trimethylammonium chloride (CCC) and succinic acid-2,2-dimethylhydrazide (B-9) did not affect cambial activity or wood formation in Malus sp. (147).

A mixture of IAA and GA_3 was synergistic in promoting cambial activity and xylem development in Acer pseudoplatanus, Populus nigra v. italica, and Fraxinus excelsior (208). In a more extensive investigation, Digby and Wareing (45) studied the effects of several concentrations of IAA + GA_3 on xylem and phloem development in Populus robusta. Maximum production of fully lignified xylem occurred in response to 100 ppm GA_3 + 500 ppm IAA. When IAA levels were lowered, GA_3 promoted phloem development. Macerations of the xylem revealed that short cambial derivatives were formed in the absence of IAA. In the presence of IAA, GA_3 clearly promoted the elongation of fibers, but not of vessel elements. GA_3 also was synergistic with IAA in promoting cambial activity in Fraxinus excelsior, although high IAA levels were inhibitory to this synergism (46). Zakrzewski (242) similarly observed that IAA + GA_3 had a marked synergistic effect upon cambial activity in *Quercus robur*. In addition, he found that the number of vessels decreased with increasing concentration of GA_3 in the presence of IAA and K; neither GA_3 alone nor together with K stimulated vessel formation. Lachaud (87) also found that IAA + GA_3 was more effective than IAA alone in promoting cambial activity and wood formation in *Fagus sylvatica*. When applied one day after debudding, IAA + GA_3 promoted normal earlywood formation. However, delaying the application of this PGR combination for one month after debudding resulted in the production of an unusual earlywood containing only short xylem elements, similar to those observed in other investigations (45, 84).

Scurfield (180) reported that the application of GA_3 to apices and leaves of *Eucalyptus* spp. increased stem height, diameter, and dry weight, but decreased the dry weight of leaves and roots. Total dry weight of treated plants was the same as nontreated controls, thus GA_3 acted by altering assimilate partitioning in favor of stem development. Anatomical observation revealed that GA_3 significantly increased xylem cross-sectional area and fiber and vessel element length, and decreased the cross-sectional area of phloem and periderm. In contrast, spraying GA_3 over several weeks did not increase the length of xylem fibers in *Populus tremuloides* Michx. seedlings (48). Scurfield (180) also observed that seedlings of six other hardwood species responded similarly to GA_3 treatment in allocating increased assimilate to main-stem development. In agreement, Wang et al. (205) found that supplying paclobutrazol, an inhibitor of gibberellin biosynthesis, in nutrient solution to *Malus domestica* Borkh cv. 'York Imperial' seedlings, decreased assimilate partitioning to the shoots and more than doubled that to the roots, resulting in increased root growth and initiation. Foliar application of GA_3 reversed many of the effects induced by paclobutrazol.

In summary, work with GA_3 indicates that it can influence cambial activity and both xylem and phloem development, depending upon the auxin/gibberellin ratio. GA_3 promotes the differentiation and elongation of fibers, but appears to have negative effects on the differentiation of vessel elements and on secondary wall deposition in developing xylem. Gibberellins appear to be important in controlling the amount of assimilate partitioned into stem wood formation.

4.2.4. *Ethylene*. Limited evidence suggests that cambial activity is regulated by ethylene levels. Overnight exposure of nonchilled, dormant *Fraxinus* shoots to ethylene chlorohydrin, a source of ethylene, effectively substituted for a cambial chilling requirement (60). The cambium of ethylene-exposed IAA-treated shoots reactivated fully, but neither ethylene nor IAA alone were similarly effective. A close relationship was found between levels of endogenous ethylene and cambial dormancy in *Juglans nigra* L. and *Prunus serotina* Ehrh. (Nelson unpublished, cited in (125)).

Several findings imply a role for ethylene in regulating wood formation. Robitaille and Leopold (159) observed that exogenous ethylene inhibited shoot elongation and stimulated diameter growth in *Malus domestica* Borkh. cv. Winesap. Applying ethrel to 2-year-old shoots of *Salix fragilis* resulted in localized stem swelling and wood formation at the application site (141). In contrast, application of the ethylene generator ethyl hydrogen-1-propyl phosphonate to stems of *Morus alba* increased tracheid frequency, decreased tracheid diameter, and had the overall effect of inhibiting diameter growth (182). Foliar applications of this compound to *Ulmus americana* seedlings resulted in the development of large-diameter multiseriate rays and tangential bands of cytoplasmically dense, xylem axial parenchyma containing numerous starch grains (44). Vessel element length and tracheid radial diameter were markedly reduced, and the number of tracheids per tracheid grouping was increased. In Salix fragilis, high ethrel levels induced the formation of xylem that contained conspicuous crystalline deposits (141). The fibers appeared to be highly lignified, were small, and exhibited transverse septations. Ethrel promoted greatly enlarged kino-vein formation in Eucalyptus regnans F. Meull. (127). Hillis (69) observed that kino-vein formation in Eucalyptus spp. was associated with a doubling of ethylene levels in the cambial region.

Nelson and Hillis (126) suggested that there is an association between increased endogenous ethylene levels and tension wood formation in Eucalyptus gomphocephala A.DC. Bending of branches was reported to increase ethylene levels in Pyrus malus L. and Prunus persica (99).

Ethylene inhibited basipetal transport of IAA in Carya illinoensis (Wargenh) K. Koch. (227) and promoted callose plugging of sieve plates in Salix viminilis (201).

In summary, the limited evidence available suggests that ethylene probably plays a role in breaking cambial dormancy, regulating diameter growth, and controlling the anatomical and chemical nature of developing xylem.

4.2.5. Cytokinins. Kinetin stimulated cambial activity in Aesculus hippocastanum (27), but not in Robinia pseudoacacia (84). Zakrzewski (242) found that K could increase or reduce the stimulation of cambial activity caused by IAA + GA_3 in Quercus robur, depending on the season, and that both Z and ZR interacted with IAA in stimulating cambial activity. In addition, he observed a synergistic effect between Z and IAA in promoting vessel development. A combination of IAA, GA_3, and K promoted more cambial activity and wood formation in Salix fragilis than did any one or two of these PGRs (158). In Malus sp., BA was ineffective in promoting cambial activity, but BA + NAA, compared with NAA, promoted more cambial activity and the xylem contained more rays and longer tracheids (146, 149).

Choveaux and van Staden (28) found that callus derived from cambium of Salix babylonica L. produced cytokinins, and that exogenous NAA inhibited this production.

More research with CKs must be conducted before their contribution to the control of cambial growth can be assessed.

5. EPILOGUE

Throughout the preparation of this review, it was tempting to incorporate the results of investigations concerned with PGR regulation of vascular development in nonforest dicots. This was not done because it would give the impression that we know much more than we really do about the regulation of stem growth and development in tree species. While it could well be that vascular development is under a common set of regulatory controls in all tracheophytes, the evidence so far gathered with forest trees suggests otherwise. Major differences in the PGR regulation of cambial activity and wood formation have been found between conifers and hardwoods, and between diffuse-porous and ring-porous hardwoods. Thus, research into PGR regulation of stem growth and development is needed with each tree species in its own right before we will begin to understand how each species makes a unique wood, and before there will be practical applications forthcoming from the basic knowledge gathered in this area.

The PGR data accumulated so far are meager and often contradictory. Hence it would be premature to extrapolate these data to a theory addressing the spatial and temporal regulation of wood formation in the whole tree.

Variability in stem diameter growth and wood quality arises as a result of variable subcellular processes. This is true whether the development is considered in terms of the degree of stem taper, the partitioning of photosynthate among shoots, roots and branches, the formation of juvenile, normal, spiral-grained or reaction wood, the extent of fiber elongation, or the weight or chemical composition of wood. All of these developments involve a host of subcellular processes e.g., cell division, primary-wall expansion, development of pits or perforation plates, deposition of cellulose, hemicellulose, and lignin polymers, microfibril orientation, production of extractives, storage of reserve materials, and protoplasmic autolysis. But each of these subcellular differentiation processes comprises many more minute biochemical and biophysical developments that ultimately are regulated by genes, or environment, or both. PGR research can be done at any level from whole trees to cell-free biochemical systems, and necessarily must be done at all levels before it will become possible to model PGR regulation of whole tree stem growth and development. It is, however, at the molecular level that we ultimately must probe if the biochemical/biophysical processes under PGR regulation are to be separated from those under some other form of regulation.

In almost all biological systems, simplistic theories on the regulation of multicellular developments are being replaced by ones of greater complexity as a result of continuing research that aims ultimately to understand the development at the level of gene expression. This is certainly true for the topic of cambial growth control in forest trees. A summary of the last 50 years of research into PGR regulation of stem growth and development must conclude that although there has been considerable research effort and numerous publications generated, nothing of practical use for forestry has yet emerged. Nevertheless, there can be little doubt that stem growth and development are under PGR regulation. Thus continuing research into PGRs will lead to genetic manipulation of trees for improved wood formation (172).

REFERENCES

1. Aloni R and Zimmermann MH (1983) The control of vessel size and density along the plant axis. A new hypothesis. Differentiation 24: 203-208
2. Aloni R and Zimmermann MH (1984) Length, width, and pattern of regenerative vessels along strips of vascular tissue. Bot Gaz (Chicago) 145: 50-54
3. Andersson B and Sandberg G (1982) Identification of endogenous N-(3-indoleacetyl)aspartic acid in Scots pine (Pinus sylvestris) by combined gas chromatography-mass spectrometry, using high-performance liquid chromatography for quantification. J Chromatogr 238: 151-156
4. Andersson B, Häggström N and Andersson K (1978) Identification of abscisic acid in shoots of Picea abies and Pinus sylvestris by combined gas chromatography-mass spectrometry. J Chromatogr 157: 303-310
5. Antoszewski R, Rangelov B, Dzięciol U and Bachmann S (1978) Influence of gamma radiation on the transport of indolyl-3-acetic acid in bark tissue from apple trees. Physiol Plant 44: 347-350
6. Balatinecz JJ and Farrar JL (1966) Pattern of renewed cambial activity in relation to exogenous auxin in detached woody shoots. Can J Bot 44: 1108-1110
7. Barker JE (1979) Growth and wood properties of Pinus radiata in relation to applied ethylene. N Z J For Sci 9: 15-19
8. Barnett JR (1971) Winter activity in the cambium of Pinus radiata. N Z J For Sci 1: 208-222

9. Barnett JR (1978) Fine structure of parenchymatous and differentiated Pinus radiata callus. Ann Bot (London) 42: 367-373
10. Bearder JR (1980) Plant hormones and other growth substances - their background, structures and occurrence. In: J MacMillan, ed. Hormonal Regulation of Development I. Encyclopedia of Plant Physiology 9: 9-112
11. Bilan MV and Kemp AK (1962) Effect of gibberellin on growth and development of loblolly pine seedlings. J For 60: 391-394
12. Bissing DR (1982) Variation in qualitative anatomical features of the xylem of selected dicotyledonous woods in relation to water availability. Bull Torrey Bot Club 109: 371-384
13. Blake J and Ferrell WK (1977) The association between soil and xylem water potential, leaf resistance, and abscisic acid content in droughted seedlings of Douglas-fir (Pseudotsuga menziesii). Physiol Plant 39: 106-109
14. Blake TJ, Pharis RP and Reid DM (1980) Ethylene, gibberellins, auxin and the apical control of branch angle in a conifer, Cupressus arizonica. Planta 148: 64-68
15. Blum W (1971) Über die experimentelle Beeinflussung der Reaktionsholzbildung bei Fichten und Pappeln. Ber Schweiz Bot Ges 80: 225-251
16. Bowen MR and Wareing PF (1969) The interchange of ^{14}C-kinetin and ^{14}C-gibberellic acid between the bark and xylem of willow. Planta 89: 108-125
17. Brenner ML (1981) Modern methods for plant growth substance analysis. Annu Rev Plant Physiol 32: 511-538
18. Brown AB and Cormack RGH (1937) Stimulation of cambial activity locally in the region of application and at a distance in relation to a wound, by means of heteroauxin. Can J Res Sect C 15: 433-441
19. Brown CL (1964) The influence of external pressure on the differentiation of cells and tissues cultured in vitro. In: MH Zimmermann, ed. The Formation of Wood in Forest Trees, pp 389-404. New York: Academic Press
20. Brown CL (1970) Physiology of wood formation in conifers. Wood Sci 3: 8-22
21. Brown CL (1971) Secondary growth. In: MH Zimmermann and CL Brown, eds. Trees: Structure and Function, pp 67-123. New York: Springer-Verlag
22. Brown CL and Wetmore RH (1959) Auxin transport in the long shoots of pine. Am J Bot 46: 586-590
23. Brown CL and Sommer HE (1975) An Atlas of Gymnosperms Cultured in Vitro. Georgia Forest Research Council, Macon, Georgia, USA
24. Brown KM and Leopold AC (1973) Ethylene and the regulation of growth in pine. Can J For Res 3: 143-145
25. Caruso JL, Smith RG, Smith LM, Cheng T-Y and Daves GD (1978) Determination of indole-3-acetic acid in Douglas fir using a deuterated analog and selected ion monitoring. Comparison of microquantities in seedlings and adult trees. Plant Physiol 62: 841-845
26. Casperson G (1965) Über endogene Faktoren der Reaktionsholzbildung. I. Wuchstoffapplikation an Kastanien Epikotylen. Planta 64: 225-240
27. Casperson G (1968) Wirkung von Wuchs und Hemmstoffen auf die Kambiumtätigkeit und Reaktionsholzbildung. Physiol Plant 21: 1312-1321
28. Choveaux NA and Van Staden J (1981) The effect of 1-naphthaleneacetic acid on the endogenous cytokinin content of aseptically cultured bark segments of Salix babylonica. Plant Cell Physiol 22: 1207-1214

29. Cleland RE (1986) The role of hormones in wall loosening and plant growth. Aust J Plant Physiol 13: 93-103
30. Cornforth JW, Milborrow BW, Ryback G and Wareing PF (1965) Identity of sycamore "dormin" with abscisin II. Nature (London) 205: 1269-1270
31. Creber GT and Chaloner WG (1984) Influence of environmental factors on the wood structure of living and fossil trees. Bot Rev 50: 357-448
32. Cronshaw J and Morey PR (1965) Induction of tensionwood by 2,3,5-tri-iodobenzoic acid. Nature (London) 205: 816-818
33. Cronshaw J and Morey PR (1968) The effect of plant growth substances on the development of tension wood in horizontally inclined stems of Acer rubrum seedlings. Protoplasma 65: 379-391
34. Crozier A, Loferski K, Zaerr JB and Morris RO (1980) Analysis of picogram quantities of indole-3-acetic acid by high performance liquid chromatography-fluorescence procedures. Planta 150: 366-370
35. Dann IR, Jerie PH and Chalmers DJ (1985) Short-term changes in cambial growth and endogenous IAA concentrations in relation to phloem girdling of peach, Prunus persica (L.) Batsch. Aust J Plant Physiol 12: 395-402
36. Dathe W, Sembdner G, Yamaguchi I and Takahashi N (1982) Gibberellins and growth inhibitors in spring bleeding sap, roots and branches of Juglans regia L. Plant Cell Physiol 23: 115-123
37. Davies JK, Jensen E, Junttila O, Rivier L and Crozier A (1985) Identification of endogenous gibberellins from Salix pentandra. Plant Physiol 78: 473-476
38. DeGroote DK and Larson PR (1984) Correlations between net auxin and secondary xylem development in young Populus deltoides. Physiol Plant 60: 459-466
39. DeMaggio AE (1966) Phloem differentiation: induced stimulation by gibberellic acid. Science 152: 370-372
40. Denne MP (1976) Wood production and structure in relation to bud activity in some softwood and hardwood species. Leiden Bot Series 3: 204-211
41. Denne MP and Dodd RS (1981) The environmental control of xylem differentiation. In: JR Barnett, ed. Xylem Cell Development, pp. 236-255. Tunbridge Wells, England: Castle House
42. Denne MP and Wilson JE (1977) Some quantitative effects of indoleacetic acid on the wood production and tracheid dimensions of Picea. Planta 134: 223-228
43. DeYoe DR and Zaerr JB (1976) Indole-3-acetic acid in Douglas fir. Analysis by gas-liquid chromatography and mass spectrometry. Plant Physiol 58: 299-303
44. Dickinson PR (1978) The effects of ethyl hydrogen 1-propylphosphonate on xylem development and differentiation in American elm. J Amer Soc Hort Sci 103: 701-704
45. Digby J and Wareing PF (1966) The effect of applied growth hormones on cambial division and the differentiation of the cambial derivatives. Ann Bot (London) 30: 539-548
46. Doley D and Leyton L (1968) Effects of growth regulating substances and water potential on the development of secondary xylem in Fraxinus. New Phytol 67: 579-594
47. Dvořák J (1961) Formation of secondary xylem in apricots (Prunus armeniaca L.) without the participation of buds. Biol Plant 3: 99-106

48. Einspahr DW and Van Buijtenen JP 1961 The influence of gibberellic acid on growth and fiber length of quaking aspen. For Sci 7: 43-51
49. Ek M, Ljungquist PO and Stenström E (1983) Indole-3-acetic acid production by mycorrhizal fungi determined by gas chromatography-mass spectrometry. New Phytol 94: 401-407
50. Evans ML (1985) The action of auxin on plant cell elongation. CRC Critical Reviews Plant Sci 2: 317-365
51. Fahn A and Zamski E (1970) The influence of pressure, wind, wounding and growth substances on the rate of resin duct formation in Pinus halepensis wood. Isr J Bot 19: 429-446
52. Farrar JL (1961) Longitudinal variation in the thickness of the annual ring. For Chron 37: 323-330
53. Fayle, DCF and Farrar JL (1965) A note on the polar transport of exogenous auxin in woody root cuttings. Can J Bot 43: 1004-1007
54. Feucht W, Khan MZ and Daniel P (1974) Abscisic acid in Prunus trees: isolation and the effect on growth of excised shoot tissue. Physiol Plant 32: 247-252
55. Field RJ (1974) The effect of endogenous gibberellins on the movement of labelled indole-3-acetic acid in willow stems. In: Bieleski RL, Ferguson AR and Cresswell MM, eds. Mechanisms of Regulation of Plant Growth. Bull Roy Soc N Z 12: 673-679
56. Field RJ and Peel AJ (1971) The metabolism and radial movement of growth regulators and herbicides in willow stems. New Phytol 70: 743-749
57. Firn RD (1986) Growth substance sensitivity: the need for clearer ideas, precise terms and purposeful experiments. Physiol Plant 67: 267-272
58. Forward DF and Nolan NJ (1961) Growth and morphogenesis in the Canadian forest species. IV. Radial growth in branches and main axis of Pinus resinosa Ait. under conditions of growth, suppression, and release. Can J Bot 39: 385-409
59. Fraser DA (1949) Production of spring wood with β-indole acetic acid (heteroauxin). Nature (London) 164: 542
60. Gouwentak CA (1941) Cambial activity as dependent on the presence of growth hormone and the non-resting condition of stems. Ned Acad Van Wetensch Proc 44: 654-663
61. Gouwentak CA and Maas AL (1940) Kambiumtätigkeit and Wuchstoff. II. Meded Landbouwhogesch Wageningen 44: 3-16
62. Gregory FG and Hancock CR (1955) The rate of transport of natural auxin in woody shoots. Ann Bot (London) 19: 451-465
63. Harris JM (1969) On the causes of spiral grain corewood of radiata pine. N Z J Bot 7: 189-213
64. Harris JM (1973) Spiral grain and xylem polarity in radiata pine: microscopy of cambial reorientation. N Z J For Sci 3: 363-378
65. Harris JM (1981) Spiral grain formation. In: JR Barnett, ed. Xylem Cell Development, pp 256-274. Tunbridge Wells, England: Castle House
66. Hautala E, Stafford A, Corse J and Barker PA (1986) Cytokinin variation in the sap of male and female Gymnocladus dioica. J Chromatogr 351: 560-565
67. Hejnowicz Z (1980) Tensional stress in the cambium and its developmental significance. Am J Bot 67: 1-5
68. Hejnowicz A and Tomaszewski M (1969) Growth regulators and wood formation in Pinus silvestris. Physiol Plant 22: 984-992
69. Hillis WE (1975) Ethylene and extraneous material formation in woody tissues. Phytochemistry 14: 2559-2562

70. Hollis CA and Tepper HB (1971) Auxin transport within intact dormant and active white ash shoots. Plant Physiol 48: 146-149
71. Hoque E (1982) Biochemical aspects of stress physiology of plants and some considerations of defense mechanisms in conifers. Eur J For Pathol 12: 280-296
72. Hoque E, Dathe W, Tesche M and Sembdner G (1983) Abscisic acid and its β-D-glucopyranosyl ester in saplings of Scots pine (Pinus sylvestris L.) in relation to water stress. Biochem Physiol Pflanz 178: 287-295
73. Horgan R (1981) Modern methods in plant hormone analysis. Progress in Phytochemistry 7: 137-170
74. Horgan R, Hewett EW, Purse JG, Horgan JM and Wareing PF (1973) Identification of a cytokinin in sycamore sap by gas chromatography - mass spectometry. Plant Sci Lett 1: 321-324
75. Horgan R, Hewett EW, Horgan JM, Purse J and Wareing PF (1975) A new cytokinin from Populus x robusta. Phytochemistry 14: 1005-1008
76. Jacquiot C (1957) Sur l'existence des facteurs d'organisation des tissus secondaires chez certains arbres forestiers. C R Acad Sci 244: 1246-1248
77. Jacquoit C (1964) Application de la technique de culture des tissus végétaux à l'étude de quelques problèmes de la physiologie de l'arbre. Ann Sci For 21: 317-473
78. Jenkins PA (1974) Influence of applied indoleacetic acid and abscisic acid on xylem cell dimensions in Pinus radiata D. Don. In: Bieleski RL, Ferguson AR and Cresswell MM, eds. Mechanisms of Regulation of Plant Growth. Bull Roy Soc N Z 12: 737-742
79. Jenkins PA and Shepherd KR (1972) Identification of abscisic acid in young stems of Pinus radiata D. Don. New Phytol 71: 501-511
80. Jensen E and Junttila O (1982) Indolyl-3-acetic acid from shoots of Salix pentandra. Physiol Plant 56: 241-244
81. Jensen E, Rivier L, Junttila O and Crozier A (1986) Identification of abscisic acid from shoots of Salix pentandra. Physiol Plant 66: 406-408
82. Kaldewey H (1984) Transport and other modes of movement of hormones (mainly auxins). In: TK Scott, ed. Hormonal Regulation of Development II. Encyclopedia of Plant Physiology 10: 80-148
83. Kennedy RW and Farrar JL (1965) Induction of tension wood with the antiauxin 2,3,5-tri-iodobenzoic acid. Nature (London) 208: 406-407
84. Kirschner H, Sachs T and Fahn A (1971) Secondary xylem reorientation as a special case of vascular tissue differentiation. Isr J Bot 20: 184-198
85. Kozlowski TT (1971) Growth and Development of Trees, I and II. New York: Academic Press
86. Kubowicz DB (1979) The possible relation between cytokinins and secondary xylem formation in Pinus silvestris. I. Seasonal correlations. Acta Soc Bot Pol 48: 295-303
87. Lachaud S (1983) Xylogénèse chez les Dictoylédones arborescentes. IV. Influence des bourgeons, de l'acide β-indolylacétique et de l'acide gibberellique sur la réactivation cambial et la xylogénèse dans les jeunes tiges de Hêtre. Can J Bot 61: 1768-1774
88. Lachaud S and Bonnemain JL (1981) Xylogénèse chez les Dicotylédones arborescentes. I. Modalités de la remise en activité du cambium et de la xylogénèse chez les Hêtres et les Chênes âgés. Can J Bot 59: 1222-1230

89. Lachaud S and Bonnemain JL (1982) Xylogénèse chez les dicotylédones arborescentes. III. Transport de l'auxine et activité cambiale dans les jeunes tiges de Hêtre. Can J Bot 60: 869-876
90. Lachaud S and Bonnemain JL (1984) Seasonal variations in the polar-transport pathways and retention sites of [^{3}H]indole-3-acetic acid in young branches of Fagus sylvatica L. Planta 161: 207-215
91. Lang GA, Early JD, Arroyave NJ, Darnell RL, Martin GC and Stutte GW (1985) Dormancy: toward a reduced, universal terminology. Hort-Science 20: 809-812
92. Larson PR (1960) A physiological consideration of the springwood summerwood transition in red pine. For Sci 6: 110-122
93. Larson PR (1962) Auxin gradients and the regulation of cambial activity. In: TT Kozlowski, ed. Tree Growth, pp 97-117. New York: Ronald Press
94. Larson PR (1963) Stem form development of forest trees. For Sci Mon 5
95. Larson PR (1964) Some indirect effects of environment on wood formation. In: MH Zimmermann, ed. The Formation of Wood in Forest Trees, pp 345-364. New York: Academic Press
96. Lavender DP and Hermann RK (1970) Regulation of the growth potential of Douglas fir seedlings during dormancy. New Phytol 69: 675-694
97. Lavender DP, Hermann RK and Zaerr JB (1970) Growth potential of Douglas fir seedlings during dormancy. In: LC Luckwill and CV Cutting, eds. Physiology of Tree Crops, pp 209-222. New York: Academic Press
98. Lee TS, Purse JG, Pryce RJ, Horgan R and Wareing PF (1981) Dihydroconiferyl alcohol- a cell division factor from Acer species. Planta 152: 571-577
99. Leopold AC, Brown KM and Emerson FH (1972) Ethylene in the wood of stressed trees. HortScience 7: 175
100. Lepp NW and Peel AJ (1971) Distribution of growth regulators and sugars by the tangetial and radial transport systems of stem segments of willow. Planta 99: 275-282
101. Letham DS, Goodwin PB and Higgins TJV (1978) Phytohormones and Related Compounds: A Comprehensive Treatise I and II. New York: Elsevier/North-Holland Biomedical Press
102. Lintilhac PM and Vesecky TB (1984) Stress-induced alignment of division plane in plant tissues grown in vitro. Nature (London) 307: 363-364
103. Lišková D (1985) Regulation of tracheids differentiation by various auxins in spruce tissue culture. Biologia (Bratislava) 40: 431-436.
104. Little CHA (1975) Inhibition of cambial activity in Abies balsamea by internal water stress: role of abscisic acid. Can J Bot 53: 3041-3050
105. Little CHA (1981) Effect of cambial dormancy state on the transport of [1-^{14}C]indol-3-ylacetic acid in Abies balsamea shoots. Can J Bot 59: 342-348
106. Little CHA and Bonga JM (1974) Rest in the cambium of Abies balsamea. Can J Bot 52: 1723-1730
107. Little CHA and Eidt DC (1968) Effects of abscisic acid on budbreak and transpiration in woody species. Nature (London) 220: 498-499
108. Little CHA and Eidt DC (1970) Relationship between transpiration and cambial activity in Abies balsamea. Can J Bot 48: 1027-1028
109. Little CHA and Loach K (1975) Effect of gibberellic acid on growth and photosynthesis in Abies balsamea. Can J Bot 53: 1805-1810

110. Little CHA and Wareing PF (1981) Control of cambial activity and dormancy in Picea sitchensis by indol-3-ylacetic and abscisic acids. Can J Bot 59: 1480-1493
111. Little CHA, Andrew DM, Silk PJ and Strunz GM (1979) Identification of cytokinins zeatin and zeatin riboside in Abies balsamea. Phytochemistry 18: 1219-1220
112. Little CHA, Heald JK and Browning G (1978) Identification and measurement of indoleacetic and abscisic acids in the cambial region of Picea sitchensis (Bong.) Carr. by combined gas chromatography-mass spectrometry. Planta 139: 133-138
113. Little CHA, Strunz GM, La France R and Bonga JM (1972) Identification of abscisic acid in Abies balsamea. Phytochemistry 11: 3535-3536
114. Lorenzi R, Horgan R and Heald JK (1975a) Gibberellins in Picea sitchensis Carriere: seasonal variation and partial characterization. Planta 126: 75-82
115. Lorenzi R, Horgan R and Heald JK (1976) Gibberellin A_9 glucosyl ester in needles of Picea sitchensis. Phytochemistry 15: 789-790
116. Lorenzi R, Horgan R and Wareing PF (1975b) Cytokinins in Picea sitchensis Carriere: identification and relation to growth. Biochem Physiol Pflanz 168: 333-339
117. McQuilkin WE (1950) Effects of some growth regulators and dressings on the healing of tree wounds. J For 48: 423-428
118. Meicenheimer RD and Larson PR (1985) Exogenous auxin and N-1-naphthylphthalamic acid effects on Populus deltoides xylogenesis. J Exp Bot 36: 320-329
119. Meyer A, Schneider G and Sembdner G (1986) Endogenous gibberellins and inhibitors in the Douglas-fir. Biol Plant 28: 52-56
120. Morey PR and Cronshaw J (1966) Induced structural changes in cambial derivatives of Ulmus americana. Protoplasma 62: 76-85
121. Morey PR and Cronshaw J (1968a) Developmental changes in the secondary xylem of Acer rubrum induced by gibberellic acid, various auxins and 2,3,5-tri-iodobenzoic acid. Protoplasma 65: 315-326
122. Morey PR and Cronshaw J (1968b) Developmental changes in the secondary xylem of Acer rubrum induced by various auxins and 2,3,5-tri-iodobenzoic acid. Protoplasma 65: 287-313
123. Morey PR and Cronshaw J (1968c) Induction of tension wood by 2,4-dinitrophenol and auxins. Protoplasma 65: 393-405
124. Nečesaný V (1958) Effect of β-indoleacetic acid on the formation of reaction wood. Phyton (Buenos Aires) 11: 117-127
125. Nelson ND and Hillis WE (1978a) Association between altitude and xylem ethylene levels in Eucalyptus pauciflora. Aust For Res 8: 69-73
126. Nelson ND and Hillis WE (1978b) Ethylene and tension wood formation in Eucalyptus gomphocephala. Wood Sci Technol 12: 309-315
127. Nelson ND and Hillis WE (1978c) Genetic and biochemical aspects of kino vein formation in Eucalyptus. I. Genetic variation in response to kino induction in E. regnans. Aust For Res 8: 75-81
128. Ng PP, Cole ALJ, Jameson PE and McWha JA (1982) Cytokinin production by ectomycorrhizal fungi. New Phytol 91: 57-62
129. Nix LE and Villiers K (1985) Tracheid differentiation in southern pines during the dormant season. Wood Fiber Sci 17: 397-403
130. Nix LE and Wodzicki TJ (1974) The radial distribution and metabolism of IAA-^{14}C in Pinus echinata stems in relation to wood formation. Can J Bot 52: 1349-1355

131. Odani K (1975a) The effects of indoleacetic acid and chilling on cambial activity of Pinus densiflora. J Jap For Soc 57: 112-116
132. Odani K (1975b) Transport of indoleacetic acid in dormant shoots of Populus deltoides. J Jap For Soc 57: 144-147
133. Odani K (1980) Effects of chilling on exogenous indole-3-acetic acid mediated wood formation and vessel diameter of white ash and poplar. Mokuzai Gakkaishi 26: 127-131
134. Odani K (1985) Indole-3-acetic acid transport in pine shoots under the stage of true dormancy. J Jap For Soc 67: 332-334
135. Odén P-C, Andersson B and Gref R (1982) Identification of gibberellin A_9 in extracts of Norway spruce [Picea abies (L.) Karst.] by combined gas chromatography-mass spectrometry. J Chromatogr 247: 133-140
136. Oserkowsky J (1942) Polar and apolar transport of auxin in woody stems. Am J Bot 29: 858-866
137. Patrick JW (1986) Hormonal control of assimilate partitioning. Plant Growth Regulator Bull 14: 7-11
138. Perry TO (1971) Dormancy of trees in winter. Science 171: 29-36
139. Pharis RP, Jenkins PA, Aoki H and Sassa T (1981) Hormonal physiology of wood growth in Pinus radiata D. Don: Effects of gibberellin A_4 and the influence of abscisic acid upon [^{3}H]gibberellin A_4 metabolism. Aust J Plant Physiol 8: 559-570
140. Phelps JE, McGinnes EA, Pieniążek J, Saniewski M and Smoliński M (1975) A scanning electron microscope study of the structures of normal, tension and morphactin IT 3456 induced wood in Aesculus hippocastanum. Bull Acad Pol Sci 23: 495-497
141. Phelps JE, McGinnes EA, Saniewski M, Pieniażek J and Smoliński M (1980) Some anatomical observations on the effect of morphactin IT 3456 and ethrel on wood formation in Salix fragilis L. Int Assoc Wood Anat Bull 1: 76-82
142. Phelps JE, McGinnes EA, Smoliński M, Saniewski M and Pieniążek J (1977) A note on the formation of compression wood induced by morphactin IT 3456 in Thuja shoots. Wood Fiber 8: 223-227
143. Phelps JE, Nelson ND, Rietveld WJ and McGinnes EA (1983) Rates of ethylene production by parenchyma cells in black walnut sapwood. Wood Fiber Sci 15: 23-27
144. Phelps JE, Saniewski M, Smoliński M, Pieniążek J and McGinnes EA (1974) A note on the structure of morphactin-induced wood in two coniferous species. Wood Fiber 6: 13-17
145. Philipson JJ and Coutts MP (1980) Effects of growth hormone application on the secondary growth of roots and stems in Picea sitchensis (Bong.) Carr. Ann Bot (London) 46: 747-755
146. Pieniążek J and Jankiewicz LS (1966) Combined effect of napthalene acetic acid and 6-benzylamino-purine on bud development and on initiation of cambial activity in dormant apple seedlings. Bull Acad Pol Sci 14: 805-808
147. Pieniążek J and Saniewski M (1968a) Hormonal control of cambial activity and xylem differentiation in apple shoots. Acta Agrobot 21: 113-129
148. Pieniążek J and Saniewski M (1968b) The effect of some growth regulators on starch depletion in apple shoots. Bull Acad Pol Sci 16: 39-46
149. Pieniążek J, Smoliński M and Saniewski M (1970) Induced structural changes in anatomy of apple shoots after treatment with morphactin IT 3456 and other growth regulators (NAA, GA, BA). Acta Agrobot 23: 387-396

150. Pomerleau R (1966) Relation of early cambial activity in white elm and infection by Ceratocystis ulmi (Buis) C. Moreau. Can J Bot 44: 109-111
151. Porandowski J, Rakowski K and Wodzicki TJ (1982) Apical control of xylem formation in the pine stem. II. Responses of differentiating tracheids. Acta Soc Bot Pol 51: 203-214
152. Purse JG, Horgan R, Horgan JM and Wareing PF (1976) Cytokinins of sycamore spring sap. Planta 132: 1-8
152A. Queiroz-Claret C, Girard Y, Girard B and Queiroz O (1985) Spontaneous long-period oscillations in the catalytic capacity of enzymes in solution. J Interdiscipl Cycle Res 16:1-9
153. Reeve Dr and Crozier A (1980) Quantitative analysis of plant hormones. In: J MacMillan, ed. Hormonal Regulation of Development I. Encyclopedia of Plant Physiology 9: 203-280
154. Reinders-Gouwentak CA (1965) Physiology of the cambium and other secondary meristems of the shoot. In: W Ruhland, ed. Encyclopedia of Plant Physiology 15 (1): 1077-1105. New York: Springer-Verlag
155. Riding RT and Little CHA (1984) Anatomy and histochemistry of Abies balsamea cambial zone cells during the onset and breaking of dormancy. Can J Bot 62: 2570-2579
156. Riding RT and Little CHA (1986) Histochemistry of the dormant vascular cambium of Abies balsamea: changes associated with tree age and crown position. Can J Bot 64: 2082-2087
157. Riov J and Gottlieb HE (1980) Metabolism of auxin in pine tissues: indole-3-acetic acid conjugation. Physiol Plant 50: 347-352
158. Robards AW, Davidson E and Kidwai P (1969) Short-term effects of some chemicals on cambial activity. J Exp Bot 20: 912-920
159. Robitaille HA and Leopold AC (1974) Ethylene and the regulation of apple stem growth under stress. Physiol Plant 32: 301-304
160. Robnett WE and Morey PR (1973) Wood formation in Prosopis: effect of 2,4-D, 2,4,5-T, and TIBA. Am J Bot 60: 745-754
161. Romberger JA (1963) Meristems, growth, and development in woody plants. US Dep Agric For Serv Tech Bull 1293
162. Ross SD, Pharis RP and Binder WD (1983) Growth regulators and conifers: their physiology and potential uses in forestry. In: LG Nickell, ed. Plant Growth Regulating Chemicals II, pp 35-78. Boca Raton, FL, USA: CRC Press
163. Saks Y and Aloni R (1985) Polar gradients of tracheid number and diameter during primary and secondary xylem development in young seedlings of Pinus pinea L. Ann Bot (London) 56: 771-778
164. Sandberg G (1984) Biosynthesis and metabolism of indole-3-ethanol and indole-3-acetic acid by Pinus sylvestris L. needles. Planta 161: 398-403
165. Sandberg G and Ericsson A (1987) Seasonal changes of indole-3-acetic acid concentrations in shoots and living stem bark of Scots pine trees and effects of pruning. Tree Physiol: submitted
166. Sandberg G and Hällgren J-E (1985) Catabolism of 3-indole acetic acid in protoplasts from etiolated seedlings of Scots pine (Pinus sylvestris L.). Plant Cell Rep 4: 100-103
167. Sandberg G and Odén P-C (1982) Effects of a short-day treatment on pool size, synthesis, degradation and transport of 3-indole-acetic acid in Scots pine (Pinus sylvestris L.) seedlings. Physiol Plant 55: 309-314
168. Sandberg G, Andersson B and Dunberg A (1981) Identification of 3-indolacectic acid in Pinus sylvestris L. by gas chromatography-mass spectrometry, and quantitative analysis by ion-pair reversed-

phase liquid chromatography with spectrofluorimetric detection. J Chromatogr 205: 125-137

169. Sandberg G, Jensen E and Crozier A (1984) Analysis of 3-indole carboxylic acid in Pinus sylvestris needles. Phytochemistry 23: 99-102

170. Sandberg G, Odén P-C and Dunberg A (1982) Population variation and diurnal changes in the content of indole-3-acetic acid of pine seedlings (Pinus sylvestris L.) grown in a controlled environment. Physiol Plant 54: 375-380

171. Savidge RA (1983) The role of plant hormones in higher plant cellular differentiation. II. Experiments with the vascular cambium, and sclereid and tracheid differentiation in the pine, Pinus contorta. Histochem J 15: 447-466

172. Savidge RA (1985) Prospects for manipulating vascular-cambium productivity and xylem-cell differentiation. In: MGR Cannell and JE Jackson, eds. Attributes of Trees as Crop Plants, pp 208-227. Instit Terrestrial Ecol, Monks Wood Exper Sta, Abbots Ripton, Huntingdon, PE17 2LS, England

172A. Savidge RA (1987) Dihydroconiferyl alcohol in developing xylem of Pinus contorta. Phytochemistry 26: 93-94

173. Savidge RA and Farrar JL (1984) Cellular adjustments in the vascular cambium leading to spiral grain formation in conifers. Can J Bot 62: 2872-2879

174. Savidge RA and Wareing PF (1981a) A tracheid-differentiation factor from pine needles. Planta 153: 395-404

175. Savidge RA and Wareing PF (1981b) Plant-growth regulators and the differentiation of vascular elements. In: JR Barnett, ed. Xylem Cell Development, pp 192-235. Tunbridge Wells, England: Castle House

176. Savidge RA and Wareing PF (1982) Apparent auxin production and transport during winter in the nongrowing pine tree. Can J Bot 60: 681-691

177. Savidge RA and Wareing PF (1984) Seasonal cambial activity and xylem development in Pinus contorta in relation to endogenous indol-3-yl-acetic and (S)-abscisic acid levels. Can J For Res 14: 676-682

178. Savidge RA, Heald JK and Wareing PF (1982) Non-uniform distribution and seasonal variation of endogenous indol-3yl-acetic acid in the cambial region of Pinus contorta Dougl. Planta 155: 89-92

179. Savidge RA, Mutumba GMC, Heald JK and Wareing PF (1983) Gas chromatography-mass spectroscopy identification of 1-aminocyclopropane-1-carboxylic acid in compressionwood vascular cambium of Pinus contorta Dougl. Plant Physiol 71: 434-436

180. Scurfield G (1962) Effects of gibberellic acid on woody perennials with special reference to species of Eucalyptus. For Sci 8: 168-179

181. Sembdner G, Gross D, Liebisch H-W and Schneider G (1980) Biosynthesis and metabolism of plant hormones. In: J MacMillan, ed. Hormonal Regulation of Development I. Encylopedia of Plant Physiology 9: 281-444

182. Sharma HK, Sharma DD and Paliwal GS (1979) Effect of chlorflurenol and FMC 10637 on cambial activity and xylem differentiation in Morus alba. Phytomorphology 29: 53-56

183. Sharma HK, Sharma DD and Paliwal GS (1980) Induction of sieve-tube elements in the ray system of Morus alba L. Biol Plant 22: 152-153

184. Shaybany B and Martin GC (1977) Abscisic acid identification and its quantitation in leaves of Juglans seedlings during waterlogging. J Am Soc Hortic Sci 102: 300-302

185. Sheldrake AR (1973) The production of hormones in higher plants. Biol Rev 48: 509-559
186. Sheriff DW (1983) Control by indole-3-acetic acid of wood production in Pinus radiata D. Don segments in culture. Aust J Plant Physiol 10: 131-135
187. Smoliński M, Saniewski M and Pieniążek J (1972) The effect of morphactin IT 3456 on cambial activity and wood differentiation in Picea excelsa. Bull Acad Pol Sci 20: 431-435
188. Smoliński M, Saniewski M, Pieniążek J, Phelps JE and McGinnes EA (1980) Anatomical observations on the influence of morphactin on wood formation of Carpinus betulus L. and Syringa vulgaris L. shoots. Acta Agrobot 33: 103-107
189. Söding H (1961) Meristeme. Gewebekulturen. In: W Ruhland, ed. Encyclopedia of Plant Physiology 14: 812-828. New York: Springer-Verlag.
190. Starbuck CJ and Phelps JE (1986) Induction of compression wood in rooted cuttings of Pseudotsuga menziesii (Mirb.) Franco by indole-3-acetic acid. Int Assoc Wood Anat Bull 7: 13-16
191. Stiebeling B, Pharis RP, Taylor JS, Wodzicki T, Abe H and Little CHA (1985) An analysis of hormones from the cambial region of Scots pine. 12th Intl Conf on Plant Growth Substances, Heidelburg, FRG
192. Studhalter RA, Glock WS and Agerter SR (1963) Tree growth. Bot Rev 29: 245-365
193. Sundberg B, Little CHA, Riding RT and Sandberg G (1987). Levels of indol-3-ylacetic acid in the vascular cambium region of Abies balsamea trees during the activity-rest-quiescence transition. Physiol Plant: submitted
194. Sundberg B, Sandberg G and Jensen E (1985a) Catabolism of indole-3-acetic acid to indole-3-methanol in a crude enzyme extract and in protoplasts from Scots pine (Pinus sylvestris). Physiol Plant 64: 438-444
195. Sundberg B, Sandberg G and Jensen E (1985b) Identification and quantification of indole-3-methanol in etiolated seedlings of Scots pine (Pinus sylvestris L.). Plant Physiol 77: 952-955
196. Taylor JS, Koshioka M, Pharis RP and Sweet GB (1984) Changes in cytokinins and gibberellin-like substances in Pinus radiata buds during lateral shoot initation and the characterization of ribosyl zeatin and a novel ribosyl zeatin glycoside. Plant Physiol 74: 626-631
197. Telewski FW and Jaffe MJ (1986a) Thigmomorphogenesis: field and laboratory studies of Abies fraseri in response to wind or mechanical perturbation. Physiol Plant 66: 211-218
198. Telewski FW and Jaffe MJ (1986b) Thigmomorphogenesis: anatomical, morphological and mechanical analysis of genetically different sibs of Pinus taeda in response to mechanical perturbation. Physiol Plant 66: 219-226
199. Telewski FW and Jaffe MJ (1986c) Thigmomorphogenesis: the role of ethylene in the response of Pinus taeda and Abies fraseri to mechanical perturbation. Physiol Plant 66: 227-233
200. Telewski FW, Wakefield AH and Jaffe MJ (1983) Computer-assisted image analysis of tissues of Ethrel-treated Pinus taeda seedlings. Plant Physiol 72: 177-181
201. Thomas B and Hall MA (1975) The effect of growth regulators on wound-stimulated callose formation in Salix viminalis L. Plant Sci Lett 4: 9-15

202. Trewavas AJ and Cleland RE (1983) Is plant development regulated by changes in the concentration of growth substances or by changes in the sensitivity to growth substances? Trends Biochem Sci 7: 354-357
203. van Staden J and Choveaux NA (1980) Cytokinins in internodal stem segments of Salix babylonica. Z Pflanzenphysiol 96: 153-161
204. Venverloo CJ (1969) The lignin of Populus nigra L. cv. 'Italica'. A comparative study of the lignified structures in tissue cultures and the tissues of the tree. Acta Bot Neerl 18: 241-314
205. Wang SY, Byun JK and Steffens GL (1985) Controlling plant growth via the gibberellin biosynthesis system. II. Biochemical and physiological alterations in apple seedlings. Physiol Plant 63: 169-175
206. Wardrop AB and Davies GW (1964) The nature of reaction wood. VIII. The structure and differentiation of compression wood. Aust J Bot 12: 24-38
207. Wareing PF (1951) Growth studies in woody species. IV. The initiation of cambial activity in ring-porous species. Physiol Plant 4: 546-562
208. Wareing PF (1958a) Interaction between indole-acetic acid and gibberellic acid in cambial activity. Nature (London) 181: 1745-1746
209. Wareing PF (1958b) The physiology of cambial activity. J Inst Wood Sci 1: 34-42
210. Wareing PF and Roberts DL (1956) Photoperiodic control of cambial activity in Robinia pseudacacia L. New Phytol 55: 356-368
211. Wareing PF, Hanney CEA and Digby J (1964) The role of endogenous hormones in cambial activity and xylem differentiation. In: MH Zimmermann, ed. The Formation of Wood in Forest Trees, pp 323-344. New York: Academic Press
212. Webber JE, Laver ML, Zaerr JB and Lavender DP (1979) Seasonal variation of abscisic acid in the dormant shoots of Douglas-fir. Can J Bot 57: 534-538
213. Webber JE, Ross SD, Pharis RP and Owens JN (1985) Interaction between gibberellin $A_{4/7}$ and root-pruning on the reproductive and vegetative process in Douglas-fir. II. Effects on shoot elongation and its relationship to flowering. Can J For Res 15: 348-353
214. Weiler EW (1984) Immunoassay of plant growth regulators. Annu Rev Plant Physiol 35: 85-95
215. Weiler EW and Ziegler H (1981) Determination of phytohormones in phloem exudate from tree species by radioimmunoassay. Planta 152: 168-170
216. Wershing HF and Bailey IW (1942) Seedlings as experimental material in the study of "redwood" in conifers. J For 40: 411-414
217. Whitmore FW (1968) Auxin and cell wall formation in the cambial derivatives of pine and cottonwood. For Sci 14: 197-205
218. Wilson BF and Archer RR (1977) Reaction wood induction and mechanical action. Annu Rev Plant Physiol 28: 23-43
219. Wilson BF, Wodzicki TJ and Zahner R (1966) Differentiation of cambial derivatives: proposed terminology. For Sci 12: 438-440
220. Wodzicki TJ and Wodzicki AB (1973) Auxin stimulation of cambial activity in Pinus silvestris. II. Dependence upon basipetal transport. Physiol Plant 29: 288-292
221. Wodzicki TJ and Wodzicki AB (1980) Seasonal abscisic acid accumulation in stem cambial region of Pinus silvestris, and its contribution to the hypothesis of a late-wood control system in conifers. Physiol Plant 48: 443-447

222. Wodzicki TJ and Wodzicki AB (1981) Modulation of the oscillatory system involved in polar transport of auxin by other phytohormones. Physiol Plant 53: 176-180
223. Wodzicki TJ and Zajączkowski S (1974) Effect of auxin on xylem tracheids differentiation in decapitated stems of Pinus silvestris L. and its interaction with some vitamins and growth regulators. Acta Soc Bot Pol 43: 129-148
224. Wodzicki TJ, Wodzicki AB and Zajączkowski S (1979) Hormonal modulation of the oscillatory system involved in polar transport of auxin. Physiol Plant 46: 97-100
225. Wodzicki TJ, Hiroshi A, Wodzicki AB, Pharis RP and Cohen JD (1987) Investigations on the nature of the auxin-wave in the cambial region of pine stems. Plant Physiol: in press
226. Wodzicki TJ, Rakowski K, Starck Z, Porandowski J and Zajączkowski S (1982) Apical control of xylem formation in the pine stem. I. Auxin effects and distribution of assimilates. Acta Soc Bot Pol 51: 187-201
227. Wood BW (1985) Effect of ethephon on IAA transport, IAA conjugation, and antidotal action of NAA in relation to leaf abscission of pecan. J Am Soc Hortic Sci 110: 340-343
228. Worrall JG (1971) Absence of "rest" in the cambium of Douglas-fir. Can J For Res 1: 84-89
229. Worrall JG (1980) The impact of environment on cambial growth. In: CHA Little, ed. Control of Shoot Growth in Trees, pp 127-142. Can For Serv-Marit, Fredericton, N B, Canada.
230. Yamaguchi K, Itoh T and Shimaji K (1980) Compression wood induced by 1-N-Naphthylphthalamic acid (NPA), an IAA transport inhibitor. Wood Sci Technol 14: 181-185
231. Yamaguchi K, Shimaji K and Itoh T (1983) Simultaneous inhibition and induction of compression wood formation by morphactin in artifically inclined stems of Japanese larch (Larix leptolepis Gordon). Wood Sci Technol 17: 81-89
232. Yang SF (1980) Regulation of ethylene biosynthesis. HortScience 15: 238-243
233. Yokota T and Takahashi N (1980) Cytokinins in shoots of the chestnut tree. Phytochemistry 19: 2367-2369
234. Yokota T, Murofushi N and Takahashi N (1980) Extraction, purification, and identification. In: J MacMillan, ed. Hormonal Regulation of Development I. Encycopedia of Plant Physiology 9: 113-201
235. Zabkiewicz JA and Steele KD (1974) Root-promoting activity of P. radiata bud extracts. In: Bieleski RL, Ferguson AR and Cresswell MM, eds. Mechanisms of Regulation of Plant Growth. Bull Roy Soc N Z 12: 687-692
236. Zaerr JB (1969) Transport gradient of indoleacetic acid in pine seedlings. Physiol Plant 21: 1265-1269
237. Zagórska-Marek B and Little CHA (1986) Control of fusiform initial orientation in the vascular cambium of Abies balsamea stems by indol-3-ylacetic acid. Can J Bot 64: 1120-1128
238. Zajączkowski S (1973) Auxin stimulation of cambial activity in Pinus silvestris. I. The differential cambial response. Physiol Plant 29: 281-287
239. Zajączkowski S and Romberger JA (1978) Polarity of xylem formation in isolated stem segments of Pinus silvestris. Physiol Plant 44: 175-180
240. Zajączkowski S and Wodzicki TJ (1978) On the question of stem polarity with respect to auxin transport. Physiol Plant 44: 122-126

241. Zajaczkowski S, Wodzicki TJ and Romberger JA (1984) Auxin waves and plant morphogenesis. In: TK Scott, ed. Hormonal Regulation of Development II. Encyclopedia of Plant Physiology 10: 244-262
242. Zakrzewski J (1983) Hormonal control of cambial activity and vessel differentiation in Quercus robur. Physiol Plant 57: 537-542
243. Zamski E and Wareing PF (1974) Vertical and radial movement of auxin in young sycamore plants. New Phytol 73: 61-69

Plant Growth Regulation 6: 171–191

8. THE ROLE OF PLANT GROWTH REGULATORS IN DORMANCY IN FOREST TREES

DENIS P. LAVENDER AND SALIM N. SILIM

Forest Sciences Department, 191-2357 Main Mall, The University of British Columbia, Vancouver, B. C., V6T 1W5

ABSTRACT. The role of plant growth regulators, i.e. indole-3-acetic acid, abscisic acid, gibberellins, cytokinins, and ethylene, in the initiation, maintenance, and breaking of dormancy in perennial, woody, temperate plants is reviewed. Particular topics discussed include: a) roles of individual compounds and antagonistic or synergistic effects of several growth regulators acting simultaneously; b) problems of compound isolation and identification; and c) hypotheses for the "modus operandi" of plant growth regulators.

1. INTRODUCTION

In the normal annual cycle of growth and development, woody plants in temperate climates go through periods characterized by shoot growth and that of no growth. The latter period, when no shoot elongation takes place, is the dormant period and is normally accompanied by terminal bud formation or shoot tip abortion. Periods of apparent inactivity in other members of the plant kingdom, such as bacteria, are also referred to as dormancy, but this paper will be concerned only with dormancy in perennial woody plants.

Shoots of woody perennial plants undergo different states of dormancy (86, 97) which can be broadly grouped into two. The first occurs when extension is arrested due to unfavourable environmental conditions, i.e. moisture stress or short photoperiods, and the second type is characterized by the inability of the bud or shoot to undergo elongation even if suitable environmental conditions for growth prevail. These dormancy states in higher plants have been referred to by a wealth of terminologies. Unfortunately, because of the array and vagueness of these terminologies, the dormancy status in plants is not always clearly defined, making the interpretation of studies on the dormancy phenomenon difficult. The stage when shoot growth may resume on exposure to favourable conditions or after defoliation has been variously referred to as "quiescence," "imposed dormancy," or "correlative dormancy;" the state characterized by lack of shoot growth under favourable conditions has been termed "rest," "true dormancy," "innate dormancy," or "winter dormancy" (86, 88, 97, 102).

Saunders (89) distinguished between ecological and physiological dormancy. Ecological dormancy includes all cases where growth ceases and metabolic activities are reduced in response to a potential unfavorable environment, and is divided into "internal" and "imposed" dormancy. This is used in a broader sense regardless of whether the dormancy is induced by unfavourable environmental condition or a condition within the plant itself. Physiological dormancy, on the other hand, is used in a much narrower sense to describe a state in which the bud is not subjected to the inhibitory effects of apical dominance or control and yet growth or bud burst is restricted even if the plant is transfered to favourable growth conditions (89). This is further subdivided into "correlative" dormancy when the stimulus arises elsewhere in the plant, for example as in leaves of Betula pubescens (99) and "innate" dormancy when the stimulus is from within the buds themselves.

More recently, Lang et al. (54) introduced the terms "eco-dormancy", "ecto-dormancy" and "endo-dormancy" to describe the different types or states of dormancy. Eco-dormancy is used to describe the dormancy state regulated by environmental factors like unfavourable temperatures, nutrient deficiency and water stress. Ecto-dormancy is used to describe the dormancy phenomenon regulated by physiological factors outside the affected structure like apical dominance and inhibition by leaves in Betula, while endo-dormancy is used to describe a condition which is regulated by factors inside the affected structure and is roughly equivalent to "rest" or "innate dormancy."

Another factor which usually makes understanding of the dormancy phenomenon difficult is the apparent lack of recognition of the morphological differences in growth patterns among species, and between the adult trees endogenous to temperate zones and their seedling stages. On a morphological basis, shoot growth in trees is characterized by two general patterns of growth usually referred to as; "preformed" or "determinate" growth and "neoformed" or "indeterminate" growth (51). In preformed growth (as in many of the gymnosperms and some angiosperms), the current season's shoot is telescoped in the bud formed during the previous year. Accordingly, the extent of current shoot growth is morphologically restricted by what has been laid down in the previous season. Thus, there are numerous cases where terminal bud formation takes place when environmental conditions (including photoperiod) are still favourable (57). In neoformed growth (or "free growth" (53)), shoot elongation is not restricted to enlargement of the primodia laid down the previous season but is influenced by current environmental conditions. Examples of neoformed growth are found in conifers such as Thuja and in the angiosperm genera. Betula, Populus, and Acer (51).

Most comprehensive studies (85) of the dormancy in tree species utilize seedlings because working with mature trees is difficult. However, seedlings of virtually all tree species have different growth patterns from mature trees (31, 51). Thus, seedlings of a number of tree species may continue growth late into the season, until restricted by low temperatures or shortening day lengths, for example, whereas mature trees stop growing earlier. Furthermore, a vast majority of species express a neoformed growth pattern in the seedling stage even though they are strictly preformed in the adult phase.

"Dormancy" is often used inaccurately to describe a morphological state in which a bud is present and this may not be related to the physiological state of the apical meristem nor to the internal development of the bud. A conifer bud is termed "dormant" early in the fall, but primordia formation may continue within the bud until early winter (77). And Lavender (59) reviewed a number of papers which demonstrate sharply differing physiological states during either innate or imposed dormancy.

The induction, maintenance and release of dormancy in tree species of temperate zones are triggered by environmental factors, including: 1) photoperiod; 2) water stress; 3) temperature; 4) mineral deficiency and 5) light quality. The first three of these factors are primary determinants of tree growth; the last two are generally considered secondary under natural conditions and have been demonstrated most clearly in the laboratory. The effects of these factors have been reviewed by Nooden and Weber (74) and Lavender (57) and will be covered only briefly here. The discovery that photoperiod plays a role in dormancy of tree species dates back to the end of the last century (74), but it was not until the 1930's that it was demonstrated that short photoperiods (SD) induced dormancy, and that, in some cases, long photoperiods (LD) resulted in release from dormancy. Water stress also results in growth cessation and terminal bud formation in a number of tree species (60, 63, 57), as do deficiencies of nutrients, especially nitrogen (51). Finally, McCreary et al., (70) discuss the role of light intensity in initiating dormancy.

Temperature also plays a major role in dormancy. Vegis (97) discusses the role of temperature in the annual growth cycle in detail and introduces the scheme of a progressively narrower temperature range in which growth is possible as dormancy develops and a correspondingly widening range of growth-permitting temperatures as dormancy is released. Numerous studies have shown that exposure to low temperature, ca 5°C (chilling), is required to release bud dormancy in several tree species (11, 30, 62). But this same temperature, which is unfavorable for active growth, helps maintain the dormant condition after the chilling requirement has been met. The role of temperature in the annual cycle is yet more complex, however, as temperature which stimulates dormancy in the fall may promote resumption of growth in spring (57, 97).

2. PHYTOHORMONES AND TREE DORMANCY

The concept of hormonal regulation in tree dormancy is probably as old as the concept of plant hormones itself, as the realization that phytohormones were involved in different aspects of tree development made it logical to expect that they were involved in dormancy regulation. Studies of phytohormone relation of dormancy in trees have evolved around two major approaches. The first approach involves measuring changes in endogenous levels at the shoot apices, the leaves, in the transpirational stream, or roots with changes in dormancy state (100, 103). In the second approach plant growth regulators (PGRs) are exogenously applied to the plant in an attempt to induce or break dormancy under conditions that would ordinarily not result in the induction or release.

Although these techniques have provided virtually all the data which describe the relationship of plant growth regulators (PGRs) to dormancy, the methodology is not sufficiently precise to yield authentic casual roles for PGRs in the physiology of dormancy. Examples of deficiencies which have been identified in PGR analyses and their interpretations include: 1) collection and storage techniques for tissue samples, which did not follow strict protocols designed to minimize diurnal variation and PGR degradation, may have resulted in spurious evaluations; 2) extraction and isolation protocols may result in artifacts or degradation of PGRs; 3) present protocols for extraction of PGRs do not permit evaluation of possible compartmentalization, changes in permeability of membranes, or availability of PGR receptor sites; 4) the active form of the PGR is frequently not distinguished from the bound state (106); 5) extraction and isolation of single PGRs does not permit evaluation of synergistic or antagonistic effects of other PGRs (50); and, finally 6) failure to recognize that tissue sensitivity may play a major role in response to PGRs (94).

All of the above deficiencies were common in studies reported prior to the past decade. Since ca 1975 more rigorous techniques for isolation and identification of PGRs, such as monoclonal antibodies, high performance liquid chromatography (HPLC), gas-liquid chromatography (GLC), and mass spectrometry (MS), have largely replaced bioassays and paper or thin layer chromatographic techniques. Even so, there remain unresolved problems in isolating many members of the gibberellin complex and in selectively extracting tissues i.e. whole buds are extracted intact rather discrete meristems.

Techniques involving exogenous application of PGRs have several limitations also. There are unsolved problems with: 1) penetration into the target tissue; 2) PGR degradation prior to uptake; and 3) metabolism of the applied PGRs after uptake. There is evidence that little of the applied PGRs reach the target tissues (38, 91). And, if hormonal regulation involves compartmentalization, as appears to be the case (33, 34), it is logical to assume that entry of the applied PGR into the "right" compartment would also be regulated in response to environmental cues (92). PGRs applied during the "wrong" environment, then, may not be guided to the correct action location. Plant tissues appear to be capable of metabolizing the applied substances (38, 91) and this ability varies with developmental stage (34) and the environment during and after application (93). In interpretation of results from exogenous application of plant growth regulators one must keep in mind that where a response is obtained high levels of the PGR were applied. However, most reports of studies which utilized this empirical approach do not include data describing the levels of PGR which actually reached the target tissue.

Some studies of trees maintained under natural conditions have demonstrated correlations between levels of PGRs and the annual cycle of growth (7, 24, 103); other reports have failed to present such evidence (106). Similar results have been obtained using container grown trees under greenhouse and growth chamber conditions (7, 89). Such contradictions under controlled conditions may have resulted from the use of extreme environments, such as very long or short photoperiods (65), which could override the effects of applied PGRs or from a failure to properly identify the dormancy status of experimental material under either natural or artificial conditions.

The remainder of this paper will review the literature describing the relationship of the several major plant growth regulators, both individually and collectively, upon development and release of dormancy. The interpretation of the data presented is limited by conceptual and procedural problems described thus far. Further, while the studies discussed utilized temperate species, with presumably annual growth cycles including definite periods of dormancy, the exact stage of dormancy of the experimental material at time of treatment is frequently not noted.

2.1. Inhibitor

The hypothesis that specific inhibitory substances (β -inhibitor complex) were involved with tree dormancy was first put forward by Hemberg (36) and supported by his work with *Fraxinus excelsior* (36). He found that the activity of ether extracts from dormant buds sampled at different periods during winter showed a marked decline (as measured by the *Avena* curvature test) as the dormancy of the buds declined. It was not, however, until after the introduction of paper chromatography to plant growth substance analysis by Bennett-Clark and Kefford (9) that an inhibitory zone on the chromatogram could be identified. The eluate from this zone was later named "β-inhibitor-complex."

Until the mid 1960s the exact nature or composition of the β-inhibitor -complex was, however, not known. Phenolics were the first group of compounds to be identified in the fraction, and for a while, were considered to be a major group of plant growth inhibitors (48). However, it was through the work of Ohkuma *et al.* (75) and Cornforth *et al.* (13) that an active fraction showing the greatest inhibitory activity was isolated and the structure elucidated. This fraction, originally referred to differently as "dormin" and "abscisin II", was later named abscisic acid (ABA).

Phillips and Wareing (100), using extracts from buds and leaves of *Acer pseudoplatanus*, found seasonal changes in inhibitor activity which appeared to correspond with changes in the dormancy state. Steward and Caplin (92) had, previously, also reported that dormancy and levels of inhibitors in *Acer saccharum* decreased in parallel during winter until the inhibitory substances disappeared completely at bud break. Similar observations have been reported for *Pinus palustris* (3), and *Populus tremula* (27). Seasonal changes in inhibitor levels were also reported in the sap of *Salix viminalis* (10) and *Salix fragilis* (16). Such data may superficially relate the level of an inhibitory substance to dormancy, but they suffer from the caveats discussed in section 9.2.

A number of studies have also reported correlations in changes of levels of inhibitors with changes in photoperiod. Phillips and Wareing (82) reported an increase in inhibitor activity (apparently β-inhibitor complex) on transfer of seedlings of *A. pseudoplatanus* to SD. Kawase (46) reported higher levels of β-inhibitor complex in *Betula pubescens* and *Betula lutea* under SD. Kawase and Nitsch (47) and Eagles and Wareing (25) also found that SD resulted in an increase in β-inhibitor complex content of *B. pubescens* leaves and buds. Furthermore, Eagles and Wareing (25), reported that when crude preparations of β-inhibitor complex from leaves of *B. pubescens* were fed to leaves of seedlings of the same species maintained under 14.5 hrs of daylight, the treated plants ceased extention growth and formed resting buds. These observations were consistent with the hypothesis that bud dormancy in woody plants is correlated with higher levels of growth-inhibiting substances produced under SD conditions.

However, all observations have not supported this hypothesis. Moreover, discrepancies have been observed with a single species, at times even when the same extraction techniques were employed. Thus, Allary (2), examining the inhibitory activity in *Quercus pedunculata*, could not find significant differences in inhibitor levels from the beginning of the rest period in late summer until the following spring. And Eliasson (27), working with *Populus tremula* shoot apices, reported a marked SD-induced increase in inhibitor activity in the Rf 0.2 - 0.4 zone (in PAW) and no change at Rf 0.5 - 0.7, while Bowen and Hoad (10) using similar fractionation techniques and experimental material found the inhibitory activity of the Rf 0.5 - 0.7 zone to be markedly increased by SD. They also reported a marked reduction of gibberellin (GA)-like activity associated with exposure to short days. Later, Taylor and Dumbroff (94) could not find an increase in the β-inhibitor complex activity in *A. saccharum* that could be correlated with changes in degree of bud dormancy. Only after buds had clearly stared to grow was a decline in the β-inhibitor complex detected in *Prunus persica* (18) and other woody species.

The observed changes of the β-inhibitor complex levels with dormancy states were, in the main, based on bioassays of crudely fractionated extracts. It has since been shown that β-inhibitor complex contained other substances which alter the response in the bioassay. The other substances, including gibberellic acid (GA) and its analogues, may be found in different levels unrelated to the inhibitors, mainly ABA, being measured. Therefore the interfering substances may occur in different levels at different times and inconsistently alter the bioassay. This discovery has led to better "clean-up" in extraction and purification procedures for PGRs.

Finally, with regard to the lack of consistent correlations between levels of the β-inhibitor complex and bud break, it should be recognized that bud break *per se* is of significance only in that it is a recognizable point event. In actuality, it represents only one brief stage in shoot elongation. Furthermore, if inhibitor level is truly related to dormancy, why do the inhibitors not disappear when the chilling requirement is satisfied rather than when sufficient heat units have accumulated to permit visible growth?

2.1.1. Abscisic Acid

2.1.1.1. Abscisic acid application.

The earliest report implicating ABA in dormancy induction was probably that of El-Antably *et al.* (26), who reported that continuous foliar application of ABA resulted in growth cessation and terminal bud formation (dormant buds) in seedlings of *B. pubescens* and *A. pseudoplatanus*, and growth cessation in *Ailanthus glutinosa*, under LD conditions. Perry and Hellmers (78) also reported cessation of growth and formation of abnormal terminal buds in a northern race of *Acer rubrum* following ABA application under LD conditions. Application by dipping supplemented by daily spraying at 30 mg/l resulted in the observed response. Elongation growth resumed, however, when application was stopped. Lavender & Waring (61) demonstrated that, although uptake of authentic ABA by immersion of cut needles in a 100 mg/l solution resulted in reduction of growth, the effect was to reduce terminal shoot growth but not lateral shoot growth of Douglas-fir seedlings maintained under a 14 h photoperiod. ABA was apparently translocated only to the most active apical meristem, that of the terminal shoot.

However, Hocking and Hillman (40) working with seedlings of B. pubescens and Alnus glutinosa reported that (±) ABA applied by foliar immersion (up to 0.2mM) without spraying did not affect growth of the seedlings. And earlier, Hillman et al. (38) had also reported that application of ABA to roots or leaves of seedlings of the same species did not induce dormancy. Elongation growth was inhibited but no resting buds were formed under non-inductive conditions, neither was dormancy nor bud formation modified under inductive conditions. These were the same species that El-Antably et al. (26) had previously reported to have induced dormancy (bud set and growth cessation) by spraying and dipping in solutions of ABA.

Junttila (42) also reported that neither direct nor indirect application (via leaves) of ABA could mimic the effect of SD in inducing cessation of stem growth in two species of Salix. Johansen et al. (41) also reported that exogenous application of ABA could not result in a complete cessation of elongation growth in Salix pentandra, although growth rate was reduced. And Heide (35) noted that growth cessation or reduced shoot growth occured following application of ABA in lanolin paste to seedlings of Picea abies kept under 24-h photoperiod. No terminal buds were formed, however, and normal growth resumed 9 weeks later.

Since isolation and identification of ABA, more studies have however, involved examination of changes in endogenous levels with dormancy states. The advent of sophisticated analytical techniques, i.e. HPLC, GLC, MS, for phytohormone analysis made it possible to examine the changes with more confidence. The use of these techniques has, therefore, clarified the possible role of ABA in induction, and maintenance of dormancy, the data reported being overwhelmingly negative.

2.1.1.2. Endogenous ABA. levels not correlated with dormancy

Lenton et al. (66) were probably the first to indicate specifically that endogenous ABA levels do not correlate with dormancy under artificial growth conditions. Using GLC, they measured ABA contents of B. pubescens and A. pseudoplatanus and observed no increase when plants were transferred to dormancy-inducing conditions. Since then, many studies have reported similar observations. Alvim and Saunders (4) found no evidence that photoperiodically induced dormancy in S. viminalis was associated with changes in ABA levels. This was later confirmed when both the β-inhibitor and ABA levels were examined using bioassays and GLC methods (6). No evidence of photoperiodically induced changes in ABA or β-inhibitor complex levels were detected under artificial SD and LD conditions.

Phillips et al. (81) observed no effect of photoperiod on endogenous levels of ABA in A. pseudoplatanus. However, phaseic acid (PA), a metabolite of ABA, was detected during the light period but not at the end of the dark period, leading them to postulate that light induced the conversion of ABA to PA. Phillips and Hoffman (80) had also observed that emergence from dormancy in A. pseudoplatanus was associated with large quantities of PA but not with a reduction in ABA levels.

Harrison and Saunders (32) examined the levels of ABA in buds of saplings of B. pubescens under natural conditions, from early fall to the time of bud burst in spring. using GLC with isotope dilution, they observed no clear correlation between ABA levels and dormancy. The ratio of "free" to "bound" or "conjugated" (alkaline-hydrolyzable) ABA decreased smoothly as buds were released from dormancy. It should be noted, however that "release from dormancy" is equivalent to "initiation of shoot growth" in this and the reports we review next and does not reflect changes that may have occurred as the chilling requirement was satisfied. Also using GLC with isotope dilution, Phillips and Hoffman (80) examined the levels o "free" and "bound" ABA in buds of A. pseudoplatanus during dormancy but found changes only after the buds had been released from dormancy. At thi stage, a decline in the "free" ABA as a percentage of the total ABA was observed. The decline was from 58% to 40% (on a fresh-weight basis). On re-examing Harrison and Saunders' (32) data on B. pubescens and Wright's (105) on Fagus sylvatica Phillips and Hoffman (80) found that for B. pubescens the decrease was from 55% in October to 45% in January whereas the pattern was essentially the same for F. sylvatica. The small changes observed led them to question the validity of the hypothesis that glucosylation of "free" ABA plays a major role in releasing buds from dormancy.

2.1.1.3. **Endogenous levels of ABA correlated with dormancy.**

There are also studies in which a positive correlation of endogenous levels of ABA with dormancy was found. Most of these observations were fo trees growing under natural conditions. Thus, Wright (105) observed a decline in level of free ABA in F. sylvatica through winter, reaching the lowest level just before bud burst. Davison and Young (17) reported that ABA in the xylem sap of S. fragilis increased in fall and remained 10 time higher in winter than in summer months, with the lowest level occuring during bud swell. In a study of seasonal changes in ABA levels in the xylem sap of S. viminalis, Alvim et al. (5) reported that the ABA content rose during November and December and then declined to a minimum at about the time of bud break. Changes in ABA content, however, corresponded with changes in concentration of other unamed solutes and therefore appeared to be a function of the degree of dilution of the sap, i.e. there was a general metabolite movement from senscing leaves in fall and a dilution of the xylem sap during renewed water uptake in spring.

Dumbroff et al. (24) observed that ABA levels in buds and stems of A. saccharum were high in July and August, declined in October and then increased following leaf fall. This early winter peak was followed by another decline to the lowest level just prior to bud burst in early May. Bound ABA increased during overwintering to the highest value in spring. The authors, nevertheless, concluded that "although patterns of change observed in the study are indicative of a role of ABA in the overwintering process, ABA does not appear to be the primary factor".

Webber et al. (103) observed seasonal variations in ABA content in buds and needles of Pseudotsuga menziesii using bioassay measurements, GLC and GLC-MS. The concentrations were highest in needles and buds in the fall, whereas those in stems peaked in January and were lowest in February and March, just before bud burst. According to them, these results corresponded with the fulfillment of chilling requirements in the species, which occurs sometime in January (62).

A recent study utilizing artificial conditions examined the effect of simulated summer, fall, and winter on Pinus sylvestris seedlings (75). Progressive changes were reported in the levels of ABA with transfer from LD of "summer" to SD of "early fall", then to "late fall" to "winter" and then back to "summer". In shoots, increases of 40 to 50% were observed in both free and bound ABA 1 week after transfer to "early fall" (8-h photoperiod, 25°C/15°C). This increase was followed by another significant increase after transfer to "late fall" (8-h photoperiod at 15°/10°C) for 3 weeks and a further increase when the seedlings were transferred to "winter" conditions (8-h photoperiod at 4°C). When the seedlings were transfered back to the "summer" condition, the level of ABA decreased to the original "summer" value. Since environments used differed in photoperiod, temperature, and light intensity, the observed changes were not necessarily the result of photoperiodic changes alone but a combination of other environmental factors like temperature and light intensity. The data of the foregoing study strongly delineate definitive changes in ABA levels with changes in environment, but no definitive concurrent determination of the dormancy status was made on the seedlings. The seedlings were apparently able to resume growth on transfer to LD, and the authors stated that the seedlings were in a state of "correlative inhibition" and not "innate" dormancy.

The exact role of ABA in dormancy regulation of tree species and indeed woody perennial plants as a whole remains unclear. It seems unlikely that photoperiodic induction of dormancy involves only changes in endogenous ABA levels. Observations of the effects of exogenously applied ABA have not been very helpful in resolving the controversy. The weight of evidence indicates that application of ABA to woody perennials (particularly tree species) under LD conditions does not result in the induction of dormancy. There is some evidence that GA levels are affected by photoperiod (66, 71, 108). The reduced inhibitor activity under LD conditions therefore could have been a reflection of increased GA content rather than a reduction in ABA content. Or the possibility that photoperiod might alter sensitivity to ABA, as with GA's has also been largely ignored and yet there is evidence that this may happen (98). Water deficits result in large increases in endogenous ABA (15) and where studies were conducted on ABA levels with trees growing under natural conditions, this may have resulted in spurious correlations of ABA levels with changes in other environmental factors. For example, recent studies in which water deficit was eliminated (81, 83) have shown that ABA levels do not change with photoperiod. Finally, any or all of the factors which contribute to inexact estimations of PGRs outlined earlier in this paper could have contributed to the uncertainty concerning the role of ABA in dormancy.

In summary, doubts which have been expressed about the involvement of ABA in the control of dormancy (89), seem to be justified. The realization that GAs did not appear to separate well from the inhibitors in the earlier studies and the subsequent observations that promoter levels appear to correlate with dormancy raised the possibility that a balance between promoter and inhibitor levels controls dormancy (102). Moreover, Wareing and Phillips (101) have suggested that a fluctuating level of ABA may not be important; possibly the antagonism between a varying level of promoters like GA, and a constant level of ABA may be a significant factor in control of dormancy.

Although the effect of applied GAs on dormancy delay or release appears to be confined either to species that do not undergo "rest" or to plants in quiescence, the difficulty in determining of dormancy states in the studies makes it hard to clearly define the effects of applied GAs. For example, it is difficult to determine whether the observed responses are a result of dormancy release or an acceleration of already released dormancy (87). Further, interpretation of the effects of applications of GAs would be facilitated if the identity of endogenous GAs was determined. Unfortunately such data are difficult to obtain with current analytical technique.

Observations of the effects of applied GAs on dormancy led to examination of the correlations between endogenous GA levels with dormancy states. Unfortunately, most of these studies used bioassays, and although results from such tests give good provisional information, they do not establish the identity or purity of the GA. Numerous investigators report that endogenous GA-like activity increases in shoot tissues before or during bud outgrowth and decreases during the onset of dormancy. Eagles and Wareing (25) chromatographed extracts of buds of B. pubescens collected in mid-winter (December) and spring (April) and observed changes in activity (dwarf maize leaf bioassay). A new promotive peak was observed which corresponded with bud growth in spring. GA-like activity was also reported to increase before or just as buds initiate visible growth in Platanus (25), B. pubescens (96) and Pseudotsuga menziesii (64). An increase in GA-activity was also observed in buds of Picea sitchensis during bud break (68). A decline in GA-activity in shoot tissues of B. pubescens (20) and Picea sitchensis (68) and in phloem exudate of willow (39) during the onset of dormancy has also been reported.

More recently, Juntilla (45) reported that apical growth cessation and formation of a shoot tip abscission zone in Salix, under natural conditions, were preceded by a marked decrease in endogenous GA-like compounds. Surprisingly, these changes occurred at a time when changes in the environmental conditions were very minimal at the locality. Using more specific techniques (GLC-MS determinations), Dathe et al. (14) identified GA_3 in the bleeding sap of B. pubescens and found that the level was very low during dormancy and increased with approach to bud burst. Short-day conditions are known to induce dormancy in seedlings of a variety of tree species. And, under SD-conditions, GA-activity at the shoot tips of Platanus (66) and phloem sap of S. viminalis (39) decreased.

Results of other studies did not find GA's to decrease under SD and increase with approach to bud burst. Langrova and Sladky (55) observed a decrease in GA-activity from terminal buds of Juglans regia in early summer, well before the development of dormancy. The GA-like activity in Populus balsamifera does not increase until after dormancy is broken, on a per bud basis but not on a fresh weight basis (8). In P. sitchensis GA-activity was reported as that of two compounds. The first, a polar GA had high levels in the fall, fell to low concentrations in winter and early spring and then increased to a maximum level in late spring in the year old foliage. The concentrations, however remained low during the growing season in buds and young needles. The second, a non-polar GA demonstrated maximum concentrations in early fall, fell to a minimum in late winter and then increased during spring and summer. The period of bud burst and active shoot elongation occurred between April and July when the non-polar gibberellin was the predominant compound. During the dormant period, from November to March, the polar GA predominated. However the maximum and minimum levels of neither compound corresponded with any phase of the trees' phenology. (68).

The evidence, therefore, which suggests that changes in GA levels may play a role in dormancy regulation is weak. And the relationship remains unclear and varies with species. The mechanism triggering some of the observed changes in GA-like substances under natural conditions seems obscure. It has been shown in spinach, however, that photoperiodic control of elongation is mediated by GA (108). In this case, photoperiod seems to affect the GA turnover rate and not total GA content (72). If, then, the role of GA in dormancy is mediated by the flux of this PGR rather than absolute instantaneous levels, such a relationship may account for the contradictions reported above as flux rates are generally not determined.

The study of the role of GAs in dormancy regulation is complicated by the large number of naturally occuring GAs in higher plants (more than 60 currently identified) and the uncertainty over which ones are physiologically active. Different GAs may show different patterns of change with dormancy state (64, 68). There is therefore, a need to identify on endogenous GAs, and GA metabolism in tree buds as their dormancy status varies.

2.2.3. Cytokinins (CKs)

Many reports try to link CKs with changes in dormancy state. Unfortunately for the purpose of this review, the majority of such studies are concerned with herbacious species.

Applied N_6-benzyladenine (BA) and kinetin are reported to have resulted in bud burst in Pinus radiata and in some non-forest tree species, e.g. Malus (49). Leike (65) reported that kinetin and BA could also result in bud break but to a much lesser extent than GA's and concluded that the termination of dormancy in the eight woody species he studied is not mediated by cytokinins. Applications of BA to Pseudotsuga menziesii seedlings in mid-fall produced no discernable effect on their physiology. However, analysis of foliage treated with radioactive BA demonstrated that the seedlings metabolized the CK within a week (61).

Although not much work has been undertaken to examine seasonal changes in endogenous CKs compared to ABA or GAs in tree species, there are indications that in several plants, CK-activity increases during chilling as well as during bud burst (23, 37). Moreover, these changes appear to involve changes in species, which probably reflect changes in metabolism. The majority of the studies utilized only bioassays as methods for identifying and quantifying the CKs, only a very few utilized more advanced techniques such as HPLC. Domanski and Kozlowski (21) reported that CK-activity increased in dormant buds of Betula papyrifera and P. balsamifera following treatment with warm water or 2-cholorethanol, treatments which release these species from dormancy. The CK-activity in leaves of P. tremula and A. psuedoplatanus decreased as dormancy, as measured by speed of bud break, developed (28). And the CK which co-chromatographed with authentic zeatin, also increased in buds of P. tremula during natural termination of dormancy preceding bud burst of this species.

Under natural conditions in Wales, CK activity rose from none in December and January to detectable in February and March with maximum levels occurring near bud burst in late March in Populus x robusta, S. viminalis and A. pseudoplatanus vegetative buds and xylem sap (37). The maximum activity in the sap occurred two weeks prior to bud burst and three weeks prior to the maximum attained in the buds. It was also noted that there were probably five cytokinins and cytokinin conjugates in the buds. Chromatographic techniques demonstrated that the major cytokinin conjugate was zeatin glucoside. Trans-zeatin riboside was reported to be the major cytokinin conjugate in the xylem sap.

Two zones (fraction A and B) of CK activity were reported in P. sitchensis leaf and bud extracts (68). During periods of active growth, fraction B (later identified as trans-zeatin riboside) was predominant while A was predominant during dormancy. Fraction A increased from a November low to a maximum in March before any visible sign of bud growth was evident. As the buds began to swell, fraction A decreased dramatically.

During the overwintering period, three zones of CK-activity were observed in roots and stems of seedlings of A. saccharum (23). The activity increased gradually from November to a maximum in April, and declined sharply after bud break. A second flush in the greenhouse in April was also preceded by an upsurge in activity.

Dumbroff and Brown (23) found CK activity in Quercus robur decreased from through April and May, during leaf expansion, followed by another rise in June, two weeks before the second flush. After that, CK activity decreased progressively to reach a minimum in November and December. More recently, Smith and Schwabe (90), using paper chromatography, in seasonal sap samples of Quercus rubur, observed several zones of CK activity. One zone had a low activity in January followed by an increase in February and March, with the maximum activity observed about 3 weeks before bud break.

Although preceding observations suggest a correlation between the state of dormancy and the levels of CKs, other studies, some with the same experimental material, provide conflicting evidence. Hewett and Wareing (37) observed no increase in CK-activity in buds of P. x robusta collected from the field until the buds were actually flushing. There was also a detectable reduction in the degree of dormancy, as indicated by the number of days required for fifty percent of the buds to burst under favorable growth conditions, well before any measurable rise in activity, even in artifically chilled twigs. A zone of activity on paper chromatography (Rf 0.95, PAW 10:1:1) from bud extracts of chilled A. saccharum decreased until chilling requirements were met, and then increased sharply thereafter (94). Another zone (Rf 0.6) increased only slightly during the dormancy-breaking low-temperature exposure. Purse et al (84) detected no significant changes in activity in spring sap of A. pseudoplatanus either before or during bud break.

Thus, although a number of studies indicate that changes in CK levels (as represented by bioassay activity) correlates with dormancy states in forest tree species, the evidence is not sufficiently strong or consistent to support such a conclusion. Unfortunately, until very recently, most of the studies have used paper chromatography to isolate CK fractions from bud extracts and bioassays for identification and quantification and, as noted previously, such techniques are deficient in the precision necessary to provide truly definitive data.

Increases in activity in xylem sap are generally interpreted to mean that cytokinins synthesized in the roots are transported to the shoots to support growth. Since the emergence from dormancy is mediated by chilling of the shoots, we are left with the question of what stimulates the roots to produce cytokinins at precisely the time required to support shoot growth. Perhaps the surge in root growth, which frequently occurs just prior to and during initiation of shoot growth, stimulates production of plant growth regulators, such as cytokinins and gibberellins in the roots.

Many naturally occuring cytokinins and conjugates have been identified, and even the limited information obtained from paper chromatography indicates that different CK's and their conjugates change with dormancy states. Properly designed experiments, employing sampling schedules based on plant physiology rather than morphology, which utilize sophisticated isolation and identification techniques, such as monoclonal antibodies, high performance liquid chromatography and GLC-MS, may permit determinations of valid correlation between levels of cytokinin species and plant dormancy. The proper identification of the CKs would enable an accurate followup of specific cytokinins and hence help to identify the species which may be involved in dormancy regulation.

Studies involving exogenous applications of CKs have not produced definitive data, possibly because many topical applications employed synthetic CKs such as N_6-benzyladenine. As with the GAs, it is difficult to decide whether an observed stimulation of bud burst reflects the lifting of innate dormancy or merely the acceleration of a process already started.

3. SUMMARY AND CONCLUSION

Although we do not know which PGRs are controlling induction, depth, or termination of dormancy, the weight of the evidence is that PGRs are the most likely mechanism for such control. They are capable of modifying certain phases of dormancy unlike any other mechanism studied, including mineral nutrition, substrates, enzymes, or specific metabolic pathways. Therefore, it is quite probable that the coupling between environmental factors and mechanisms of dormancy involve PGRs. However, this paper has reviewed a wide range of reports on possible role(s) of PGRs in the initiation, maintenance, and release of dormancy in perennial temperate zone plants. The many contradictions and discrepancies in the data presented suggest that the research reported suffers from the following deficiencies:

1. A lack of a clear, physiologically valid, definition of the term "dormancy".

2. A concentration upon the effects of levels of PGRs upon plant response to the exclusion of the possible effects of changes in tissue sensitivity or of the flux of the PGR investigated.
3. Present extraction techniques yield data describing mean concentrations of PGRs for given tissue types and are not appropriate to address the possibility that PGR effects are modulated by levels in specific cellular compartments.
4. Much of the data reported reflect inadequate extraction, isolation, and identification methodologies. Further, few reports include radioisotope evaluations of the percent recovery of PGRs using given experimental techniques. Reports that do so, indicate that recoveries of endogenous hormones may be very low.
5. Lack of biochemical data that clearly define the role of PGRs in plant metabolism. Such a lack virtually prohibits establishment of a cause and effect role of PGRs in dormancy regulation.
6. Lack of precise, concurrent assessments of anatomical (primordial) changes and changes in hormone levels (given 2 & 3 above) limit cause and effect conclusions.

Finally, it is possible that the development of cold hardiness in the bud may obscure changes related solely to "dormancy" in the apical meristem and that the "conflicting results" may be the result of failure to recognize differences in pattern on timing of behaviour unique to species.

REFERENCES

1. Alden T (1971) Seasonal variation and occurence of IAA in buds of Pinus silvestris. Physiol Plant 25:54-57
2. Allary S (1961) Remarques sur le role du complexe inhibiteur B dans la dormance de bourgeons de Quercus pedunculata Ehrh. CR Acad Sci (Paris) Ser D 250:911-913
3. Allen RH (1960) Changes in acid growth substances in terminal buds. Physiol Plant 13:555-560
4. Alvim R and Saunders PF (1974) Abscisic acid and the photoinduction of dormancy in Salix viminalis. Plant Physiol Ann Suppl 69
5. Alvim R, Hewett W and Saunders PF (1976) Seasonal variations in hormone content of willow. I. Changes in abscisic acid and cytokinin activity in xylem sap. Plant Physiol 57:474-476
6. Alvim RE, Saunders PF and Barros R (1979) Abscisic acid and the photoperiodic induction of dormancy in Salix viminalis L. Plant Physiol 63:774-777
7. Bachelard EP Control of Dormancy In p 273-296 Little CHA (ed) (1980) Control of Shoot Growth in Trees, Canadian Forestry Service, Fredericton, NB
8. Bachelard EP and Wightman F (1974) Biochemical and physiological studies on dormancy release in tree buds. III. Changes in endogenous growth substances and possible mechanism of dormancy release. Can J Bot 52:1483-1489
9. Bennett-Clark TA and Kefford NP (1953) Chromatography of growth substances in plant extracts. Nature 171:645-648
10. Bowen MR and Hoad GV (1968) Inhibitor content of phloem and xylem sap obtained from willow (Salix viminalis) entering dormancy. Planta 81:64-70

11. Campbell RK and Sugano AI (1975) Phenology of bud burst in Douglas-fir related to provenance, photoperiod, chilling and flushing temperature. Bot Gaz 136:290-298
12. Corgan JN and Widmoyer FB (1971) The effects of gibberellic acid on flower differentiation, date of bloom, and flower hardiness in peach. J Amer Soc Hort Sci 96:54-57
13. Cornforth JW, Milborrow BV, Ryback G and Wareing PF (1965) Identity of sycamore "dormin" with abscisin II. Nature 205:1269-1270
14. Dathe W, Sembdner G, Kefeli VI and Vaslov PV (1978) Gibberellins, abscisic acid, and related inhibitors in branches and bleeding sap of birch (Betula pubescens Ehrh.) Bioch Physiol Pflanz 173:238-248
15. Davies WJ and Mansfield TA The role of abscisic acid in drought avoidance. In p 237-268 Addicott FT (ed) (1983) Abscisic acid New York, Pregger Publishers NY
16. Davison RM (1965) Some properties of a growth inhibitor present in the xylem sap of woody species. Aust J Biol Sci 18:475-486
17. Davison RM and Young H (1973) Abscisic acid content of xylem sap. Planta 109:95-98
18. Dennis FG and Edgerton LJ (1961) The relationship between an inhibitor and rest of peach flower buds. Proc Amer Soc Hort Sci 77:107-116
19. DeYoe DR and Zaerr JB (1976) Indole-3-acetic acid in Douglas-fir. Plant Physiol 58:299-303
20. Digby RJ and Wareing PF (1966) The relationship between endogenous hormone levels in the plant and seasonal aspects of cambial activity. Ann Bot 30:607-622
21. Domanski R and Kozlowski TT (1968) Variations in kinetin-like activity in buds of Betula and Populus during release from dormancy. Can J Bot 46:397-403
22. Dörfling K (1963) The auxin/inhibitor regime in Acer pseudoplatanus. I. The yearly course of auxins and inhibitors in buds, leaves and the cambium. Planta 60:390-412
23. Dumbroff EB and Brown DCW (1976) Cytokinins and inhibitor activity in roots and stems of sugar maple seedlings through the dormant season. Can J Bot 54:191-197
24. Dumbroff EB, Cohen OB and Webb DP (1979) Seasonal levels of abscisic acid in buds and stems of Acer saccharum. Physiol Plant 45:211-214
25. Eagles CF and Wareing PF (1964) The role of growth substances in the regulation of bud dormancy. Physiol Plant 17:697-709
26. El-Antably HMM, Wareing PF and Hillman J (1967) Some physiological responses to d, l-abscisin (dormin). Planta 73:74-90
27. Eliasson L (1969) Growth regulators in Populus tremula. I. Distribution of auxin and growth inhibitors. Physiol Plant 22:1288-1301
28. Engelbrecht L (1971) Cytokinins in buds and leaves during growth, maturity and aging. Biochem Physiol Pflanz 162:547-558
29. Farmer RE (1966) Gibberellin-induced growth of dormant sweetgum. USDA. South For Expt Stn For Serv Res Note SO-42, p.3
30. Fuchigami LH, Evert DR and Weiser CJ (1971) A translocatable cold hardiness promoter. Plant Physiol 47:164-167
31. Hanover JW (1980) Control of Tree Growth. Bio Sci 30 11:756-762
32. Harrison MA and Saunders PF (1975) The abscisic acid content of dormant birch buds. Planta 123:291-298
33. Hartung W, Gimmler H and Heilman B The compartmentation of abscisic acid (ABA), of ABA-biosynthesis, ABA-Metabolism and ABA- conjugation.

In p 325-333 Plant growth substances, (1982) Academic Press, London, New York

34. Hedden P, MacMillan J and Phinney BO (1978) The metabolism of gibberellins. Ann Rev Plant Physiol 29:149-192
35. Heide OM (1986) Effects of ABA application on cessation of shoot elongation in long-day grown Norway spruce seedlings. Tree Physiol 1:79-83
36. Hemberg T (1949) Growth inhibiting substances in terminal buds of Fraxinus. Physiol Plant 2:37-44
37. Hewett EW and Wareing PF Cytokinin changes during chilling and bud burst in woody plants. In p 693-701 Bileski RL, Ferguson AR and Cresswell MM (eds.) (1974) Mechanism of regulation of plant growth. Roy Soc NZ, Wellington
38. Hillman JR, Hocking JJ and McWha JA (1974) Abscisic acid and the regulation of dormancy in tree seedlings and lettuce fruits. In p 882-890 Plant growth substances, 1973. Hirokawa Publishing, Tokyo
39. Hoad GW and Bowen MR (1968) Evidence for gibberellin-like substances in the phloem exudate of higher plants. Planta 82:22-32
40. Hocking JJ and Hillman JR (1975) Studies on the role of abscisic acid in the initiation of bud dormancy in Alnus glutinosa and Betula pubescens. Planta 125:235-242
41. Johansen LG, Oden P-C, and Junttila O (1986) Abscisic acid and cessation of apical growth in Salix pentandra. Physiol Plant 66:409-412
42. Junttila O (1976) Apical growth cessation and shoot tip abscission in Salix. Physiol Plant 38:278-286
43. Junttila O (1978) Shoot tip abortion as affected by growth hormones and defoliation. Z. Pflanzen Physiol 87:455-462
44. Junttila O (1981) Effects of different gibberellins on elongation growth under short day conditions in seedlings of Salix pentandra. Physiol Plant 53:315-318
45. Junttila O (1982) Gibberellin-like activity in shoots of Salix pentandra as related to the elongation of growth. Can J Bot 60:1231-1234
46. Kawase M (1961) Growth substances related to dormancy in Betula. Proc Amer Soc Hort Sci 78:532-536
47. Kawase M and Nitsch JP (1958) Growth substances and the photoperiodic control of growth in Betula pubescens. Plant Physiol 33:xix
48. Kefeli VI and Kadyrov CSh (1971) Natural growth inhibitors their chemical and physiological properties. Ann Rev Plant Physiol 22:185-260
49. Kender WJ and Carpenter S (1972) Stimulation of lateral bud growth in apple trees by 6-benzylamino purine. J Amer Soc Hort Sci 97:377-380
50. Khan AA (1975) Primary, preventive and permissive roles of hormones in plant systems. Bot Rev 41:391-420
51. Kozlowski TT (1971) Growth and development of trees. Vol II Academic Press, New York
52. Kupcewicz J (1970) Gibberellin-like substances and growth inhibitors in pine trees (Pinus silvestris L.) of different ages. Sc Proc, Univ Copernicus, Torun, 23:131-138
53. Lanner RM Patterns of shoot development in Pinus and their relationship to growth potential. In p 223-243 Cannell MGR and Last FT (eds) (1976) Tree Physiology and Yield Improvement Academic Press, New York, NY

54. Lang GA, Early JD, Arroyave NJ, Darnell RL, Martin GC and Stutte GW (1985) Dormancy: towards a reduced, universal terminology. Hort Sci 20:809-812
55. Langrova V and Sladky Z (1971) The role of growth regulators in the differentiation of walnut buds (Juglans regia L.). Biol Plant 13:361-367
56. Larson PR (1960) Gibberellic acid-induced growth of dormant hardwood cuttings. Forest Sci 6:232-239
57. Lavender DP Effects of environment upon shoot growth of woody plants In p 76-106 Little CHA (ed) (1980) Control of Shoot Growth in Trees, Canadian Forestry Service, Fredericton, NB
58. Lavender DP Bud Dormancy in p 7-15 Duryea ML (ed) 1985 Evaluating Seedling Quality: principles, procedures, and predictive abilities of major tests, Oregon State University, Corvallis, Oregon
59. Lavender DP Dormancy in Hinckley TM and Lassoie JP (eds) 1987, Techniques & Approaches in Forest Tree Ecophysiology, in prep
60. Lavender DP, Ching KK and Hermann PK (1968) Effect of environment on the development of dormancy and growth of Douglas-fir seedlings. Bot Gaz 129:70-83
61. Lavender DP and Wareing PF, unpublished data, 1969
62. Lavender DP and Hermann PK (1970) Regulation of the growth potential of Douglas-fir seeedlings during dormancy. New Phytol 69:675-694
63. Lavender DP and Overton WS (1972) Thermoperiods and Soil Temperatures as They Affect Growth and Dormancy of Douglas-fir Seedlings of Different Geographic Origin. Oregon State University Forest Research Laboratory Research paper 13. Corvallis, Oregon. 26 pages.
64. Lavender DP, Sweet GB, Zearr JB and Hermann RK (1973) Spring shoot growth in Douglas-fir may be initiated by gibberellins exported from the roots. Science 182:838-839
65. Leike NH (1967) Wirkung von gibberellinsaure and kinetin auf ruhende verschiedener geholze. Flora Jena (Abst) A 158:361
66. Lenton JR, Perry VM and Saunders PF (1972) Endogenous ABA in relation to photoperiodically-induced bud dormancy. Planta 106:13-21
67. Lockhart JA and Bonner J (1957) Effect of gibberellic acid on the photoperiodically controlled growth of woody plants. Plant Physiol 32:492-499
68. Lorenzi R, Horgan R and Heald JK (1975) Gibberellin in Picea sitchensis Carrie: seasonal variation and partial characterization. Planta 126:75-82.
69. Lyr H, Hoffman G, Richter R (1970) On the chilling requirement of dormant buds of Tilia platyphyllos. Bioch Physiol Pflanz 161:133-141
70. McCreary DD, Tanaka Y and Lavender DP (1978) Regulation of Douglas-fir Seedling Growth and Hardiness by Controlling Photoperiod. Forest Science 24 (2):142-152
71. Metzger JD and Zeevart JAD (1980a) Effect of photoperiod on the levels of endogenous gibberellins in spinach as measured by combined gas chromatography-selected ion current monitoring. Plant Physiol 66:844-846
72. Metzger JD and Zeevart JAD (1980b) Photoperiodic control of gibberellin metabolism in spinach. Plant Physiol Suppl 6:77
73. Michalki L and Krysko K Seasonal changes in the dynamics of auxin and gibberellin-like substances during the development of buds of oak. Proc Third Symp. Plant Growth Reg 1970 p 147-152, Toron Poland

74. Nooden, LD and Weber JA Environmental and hormonal control of dormancy in terminal buds of plants In p 227-269 Cutter ME (ed) (1978) Dormancy and development arrest Academic Press, New York
75. Oden, PC and Dunberg A 1984. Abscisic acid in shoots and roots of scots pine (Pinus sylvestris L.) seedlings grown in controlled long-day and short-day environments. Planta 161:148-155
76. Ohkuma K, Addicott FT, Smith OE, and Thiesen CW (1965) The structure of abscisin II. Tetrahedron lett 29:2529-2535
77. Owens JN and Molder M (1976) Bud development in Sitka spruce. I Annual growth cycle of vegetative buds and shoots. Can J Bot 54:313-325
78. Perry TO and Hellmers H (1973) Effects of abscisic acid on growth and dormancy of two races of red maple. Bot Gaz 134:283-289
79. Pharis RP and Kuo CG (1977) Physiology of gibberellins in conifers. Can Jr For Res 7:299-325
80. Phillips IDJ and Hoffman A (1979) Abscisic acid, abscisic acid esters and phaseic acid in vegetative buds of Acer pseudoplatanus during emergence from winter dormancy. Planta 146:591-596
81. Phillips IDJ, Miners J and Roddick JG (1980) Effect of light and photoperiodic conditions on abscisic acid in leaves and roots of Acer pseudoplatanus L. Planta 149:118-122
82. Phillips IDJ and Wareing PF (1958) Studies in the dormancy of sycamore. I. Seasonal changes in the growth substance content of the shoots. J Expt Bot 9:350-364
83. Plancher B and Nauman WD (1978) Influence of water supply and daylength on abscisic acid content of Fragaria anannassa. Garternbauwissenschaft 43:126-136.
84. Purse JG, Horgan R, Horgan JM and Wareing PF (1976) Cytokinins of sycamore spring sap. Planta 132:1-8
85. Rodriguez A and Sanchez-Tames R (1986) Dormancy and seasonal changes in plant growth regulators in hazel buds. Physiol Plant 66:288-292.
86. Romberger JA (1963) Meristems, growth and development in woody plants. USDA For Serv Tech Bull No. 1293
87. Ross SD, Pharis RP and Binder W Growth Regulators and conifers: Their Physiology and Potential Uses in Forestry. In p 35-78. Nickell LG (ed) 1983 Plant Growth Regulating Chemicals, Vol II CRC Press, Inc Boca Raton, Fl
88. Samish RM (1954) Dormancy in woody plants. Ann Rev Plant Physiol 5:183-204
89. Saunders P Phytohormones and bud dormancy. In p 423-446 Letham DS, Goodwin PB and Higgins TJV (eds). 1978 Phytohormones and related compounds New York, NY, Elsevier
90. Smith DJ and Schwabe WW (1980) Cytokinin activity in oak (Quercus robur) with particular reference to transplanting. Physiol Plant 48:27-32
91. Sponsel, VM In vivo gibberellin biosynthesis in higher plants. In p 151-250 Crozier A (ed) (1983) The biochemistry and and physiology of gibberellins Vol 1, Preager Publishers, New York, NY
92. Steward FC And Caplin SM (1952) Investigations on growth and metabolism of plant cells. III. Evidence for growth inhibitors in certain mature tissues. Ann Bot 16:476-489
93. Stoddart JL Sites of gibberellin biosynthesis and action. p 1-56. In A. Crozier (ed) The biochemistry and physiology of gibberellins. Vol II (1983) New York, NY Preager Publishers

94. Taylor JS and Dumbroff EB (1975) Bud, root, and growth regulator activity in Acer saccharum during the dormant season. Can J Bot 53:321-331
95. Trewavas AJ (1982) Growth substance sensitivity: the limiting factor in plant development. Physiol Plant 55:60-72
96. Tumanov II, Kuzina GV and Karnikova LD (1970) Effects of gibberellins on period of dormancy and frost-resistance of plants. Fiziol Rast 17:737-745
97. Vegis A (1964) Dormancy in higher plants. Ann Rev Plant Physiol. 15:185-224
98. Vince-Prue D Photoperiod and hormones In p 308-364 Pharis RP and Reid DM (eds) 1985 Hormonal Regulation of Development, Encyclopedia of plant physiology, New Series, Vol 11, Springer Verlag, Berlin
99. Wareing PF (1954) Growth studies in woody species. VI. The locus of photoperiodic perception in relation to dormancy. Physiol plant 7:261-277
100. Wareing PF and Phillips IDJ (1958) Studies in the Dormancy of Sycamore. I. Seasonal Changes in the Growth-Substance Content of the Shoot. J Experimental Bot 9, p 350-364
101. Wareing PF and Phillips IDJ Abscisic acid in bud dormancy and apical dominance. In p 301-329 Addicott FT (ed) (1983) Abscisic Acid, Preager Publishers, New York, NY
102. Wareing PF and Saunders PF (1971) Hormones and dormancy. Ann Rev Plant Physiol 22:261-288
103. Webber JE, Laver ML, Zaerr JB and Lavender DP (1979) Seasonal variation of abscisic acid in the dormant shoots of Douglas-fir. Can J Bot 57:534-538
104. Wood BW (1983) Changes in indoleacetic acid, abscisic acid, gibberellins, and cytokinins during bud break in pecan. J Am Soc Hort Sci 108:333-338
105. Wright STC (1975) Seasonal changes in the levels of free and bound abscisic acid in black currant (Ribies nigrum) and beech (Fagus sylvatica) buds. J Exp Bot 26:161-174
106. Zaerr JB Hormonal relations In Hinckley TM and Lassoie JP (eds) 1987 Techniques and Approaches in Forest Tree Ecophysiology, in prep
107. Zaerr JB, Lavender DP and Sweet GB (1973) Growth regulators in Douglas-fir during dormancy. In Proceedings of Symposium on Dormancy in Trees, Polish Academy of Science, International Union of Forest Research Organization, Kornik, Poland
108. Zeevart JAD (1974) Endogenous gibberellins and growth responses in spinach under different photoperiods. p 1175-1181 In Plant Growth Substances, 1973. Hirokawa Publishing Co., Tokyo

Plant Growth Regulation 6: 193–215

9. STRESS PHYSIOLOGY OF FOREST TREES: THE ROLE OF PLANT GROWTH REGULATORS

J.D. JOHNSON
Assistant Professor of Tree Physiology, Department of Forestry, University of Florida, Gainesville, Florida 32611

ABSTRACT

Abiotic and biotic stresses elicit changes in normal physiology of trees. Plant growth regulators (PGR) are involved in the stress response and appear to have two roles: 1) to minimize the impact of the stress on the tree and; 2) to trigger stress resistance mechanisms. In the latter case the PGR-induced changes appear to enhance resistance to subsequent stress. This cross-adaptation to stress is important in trees.

The role of PGRs in the physiological response to the abiotic stresses of water deficit, water excess, temperature, nutrition and mechanical perturbation is discussed along with cross-adaptation in the interactions of these stresses. Disease response and defense, and plant-plant communications involve PGRs and are topics covered with respect to biotic stress. Stress leads to early senescence and abscission in trees. These processes are controlled by PGRs and are briefly discussed.

1. INTRODUCTION

Abiotic and biotic stresses impact the growth and development of forest trees throughout their lifetime. These stresses can significantly alter normal physiological processes. The extent of alteration depends on the severity and duration of the stress. Such physiological alterations have been termed strain by Levitt (43) and are categorized as elastic (reversible) and plastic (irreversible) strain. Elastic strain can be considered a normal fluctuation in physiological processes in response to a stress; in general, there is little evidence to implicate plant growth regulators (PGRs) in this response. An example of an elastic strain is a reduction in shoot growth due to low temperature (above freezing) which disappears when the temperature rises. Prolonged elastic stress, however, can eventually lead to irreversible changes and may, therefore, involve PGRs (43).

The irreversible alterations in physiological processes in response to stress, i.e. plastic strain, are where PGRs have been implicated. Itai and Benzioni (30) provided evidence for PGR control of plastic strain in response to water stress. The evidence is equally applicable to plastic strain in general. Evidence includes: 1) the concurrent response to stress throughout the plant in organs not directly exposed to the stress; 2) the developmental changes that occur in response to stress; 3) the metabolic changes that occur *in situ* that do not occur *in vitro* even though the tissues are exposed to the same degree of stress; and 4) the similarity of plant responses to different stresses. The role of PGRs in the stress response appears to be twofold. The first is to minimize the impact of the stress on the "normal" physiology and the second is to trigger physiological changes that impart stress resistance, including anatomical and morphological alterations.

Abscisic acid (ABA) and ethylene are the two classes of PGRs usually associated with plastic strain in trees caused by abiotic and biotic stress. When measured, however, the other classes of PGRs -- auxins, gibberellins and cytokinins -- have exhibited changes as well in response to stress (15,29,84). The relative changes of some or all of the PGRs may be responsible for the observed strain in trees (4,29).

The changes in PGR levels in response to stress have been demonstrated to modify a tree's subsequent response to other stresses. This concept has been termed cross-adaptation (26). Boussiba et al. (14) provide evidence for the involvement of ABA in the cross-adaptation in tobacco (*Nicotiana rustica* L.) plants. Pre-exposure of plants to various growth-reducing stresses improved their subsequent performance when exposed to either freezing temperatures or reduced root zone oxygen. Furthermore, they reported that within limits, the more severe the initial stress, the better the relative performance under subsequent stresses (14). Cross-adaptation in trees may provide a unifying concept relating plastic strain to changes in PGRs.

This chapter will examine the documented changes of PGRs in trees in response to the abiotic stresses of water, temperature, nutrition and mechanical stress, and biotic stresses including disease response and defense, and plant-plant communications. Senescence and abscission will be briefly covered since these two physiological processes usually accompany plastic strain, especially when the stress is severe or of long duration.

2. ABIOTIC STRESSES

2.1. Water

2.1.1. *Introduction*. Water has been identified as the most important environmental factor that, singly or in combination, determines the distribution of vegetation on earth (38,39). As a result of this importance, much stress physiology research in trees has focused on either the effects of water deficits or excesses.

2.1.2. *Deficits*. Water deficit in a plant develops when transpirational water loss exceeds uptake from the soil. Such deficits can occur diurnally and presumably the plant response does not involve PGRs. Such diurnal variations constitute an elastic strain (43). When soil water becomes increasingly depleted and unavailable to the tree, water deficit intensifies causing water stress.

At the onset of water stress plants respond rapidly to avoid, or at least, minimize the strain. This response is the much documented ABA-mediated stomatal closure (4,20,22,29,55,82). Milborrow (54) categorized this ABA response as fast (occurring within minutes of application of the stress) in contrast to responses he classified as slow (requiring from 20 to 30 minutes). There is some evidence for separate ABA binding sites for the two types of responses (54). ABA-mediated stomatal closure appears to be associated with ABA binding to guard cell plasma membrane, inhibiting a proton pump that exchanges protons for potassium ions which normally leads to increased turgor and stomatal opening.

Increases in leaf ABA content with increasing water stress have been correlated with stomatal closure in several tree species. Blake and Ferrell (11) reported an increase in ABA concentration in the needles of Douglas-fir (Pseudotsuga menziesii (Mirb.) Franco) seedlings that was correlated with decreases in branch conductance of water vapor and decreases in both soil and xylem water potential. Since then similar results have been reported in Douglas-fir (34,56), balsam fir (Abies balsamea (L.) Mill.) (46), Populus (Populus robusta x schied) (73) and Scots pine (Pinus sylvestris L.) (28). The response in all of these species followed the general pattern shown in Fig. 1. In those studies where changes were followed after watering (11,28) and into a second stress period (34,56), full stomatal function lagged several days behind the recovery of leaf water potential (Fig. 1). This may indicate the presence of residual ABA in the vicinity of the guard cells, stomatal impairment by ABA metabolites such as phaseic and dihydrophaseic acid (28), decreased cytokinin concentration in the leaves (30), or repair to damaged or altered plasma membrane. Given enough time, however, stomatal function will return to prestress levels leading to the suggestion that this response may be elastic.

The origin of ABA in the stomatal response appears to be de novo synthesis in chloroplasts of mesophyll cells adjacent to stomata (47); however, the release of ABA from its glucosyl ester either in guard or mesophyll cells has also been suggested (28,34). In either case the leaf water potential at which ABA increases varies with species and culture of experimental subjects. For example, in Douglas-fir needles ABA was observed to increase at leaf water potentials of -0.7 MPa (56), -1.0 MPa (11), and -2.0 MPa (34). These differences have been attributed to the triggering of ABA production, not by leaf water potential, but by leaf turgor potential (59), specifically the approach or achievement of zero turgor. Thus the differences in the threshold water potentials would be due to differing osmotic potentials encountered in the different species and experimental conditions. The genetic effects on this response has been reviewed by Aspinall (6).

flowering in trees. Although the mechanism is not known, flower initiation is related to decreased vegetative growth and increased levels of less polar gibberellins such as GA_4 and GA_7 (25,58). Water deficits apparently retard growth, leading to a shift in GA metabolism from polar to less polar GAs (58). The elevated levels of less polar GAs then induce strobilus primordia initiation in the quiescent buds (24,25).

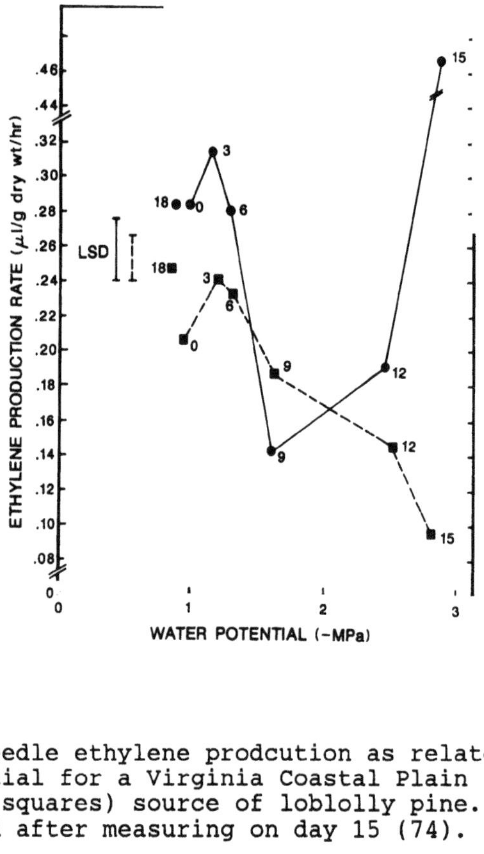

FIGURE 2. Needle ethylene prodcution as related to xylem water potential for a Virginia Coastal Plain (circles) and an east Texas (squares) source of loblolly pine. The trees were watered after measuring on day 15 (74).

Timing of the water stress has been observed to influence the sex of the strobili produced in loblolly pine (25).

2.1.3 Excesses. Too much water can be equally stressful to trees as too little (37). Excess water, or waterlogging, rapidly leads to the depletion of oxygen in and around roots. Anaerobiosis of roots leads to stomatal closure, reduced photosynthesis, epinasty, leaf chlorosis, senescence and premature abscission, reduced shoot elongation, adventitious root development, aerenchyma and hypertrophy in roots and stems (19,37,64).

The hormonal involvement in the response of plants to flooding has been reviewed recently by Reid and Bradford (64). Increased ethylene production is consistently associated with flooding in tree species. Clemens and Pearson (19) observed a doubling of the rate of ethylene

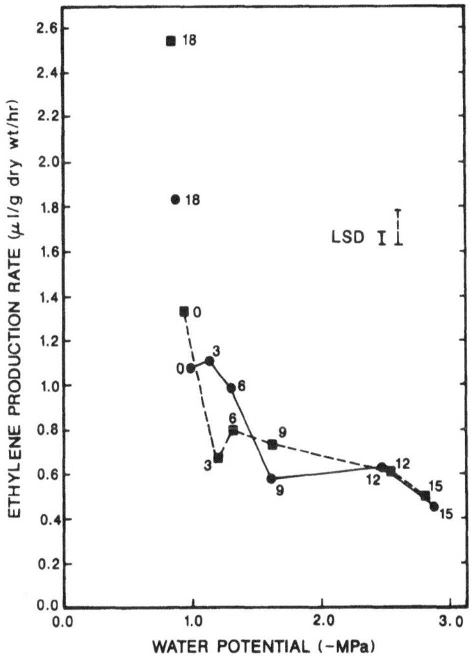

FIGURE 3. Root ethylene production as related to xylem water potential for two sources of loblolly pine. Symbols same as in Fig. 2 (74).

release from the roots and stems of *Eucalyptus robusta* Sm. within six days of flooding while production by leaves was unchanged. In flooded seedlings of *Quercus macrocarpa* Michx., ethylene production was accelerated in the lower stem leading to the formation of hypertrophied lenticels (76). A comparison of three species of *Eucalyptus* with differences in flooding tolerance suggested that tolerance was associated with ethylene production initially in roots and subsequently in the lower stem which leads to aerenchyma formation (12). The least flood tolerant of the three species exhibited no ethylene production nor any aerenchyma development. It was suggested that stem hypertrophy caused by ethylene production allowed for the elimination of potentially phytotoxic ethylene and enhanced oxygen transport to the roots (12).

A model for the role of ethylene in tree responses to flooding was proposed by Reid and Bradford (64) and is summarized in Table 1. Anaerobiosis in the roots stimulates ACC production, but its conversion to ethylene is inhibited by this condition. The ACC builds up and begins to be transported in the xylem up the stem. Near the water line oxygen is available in the stem and ACC is converted to ethylene. The ethylene stimulates the activity of cellulase which in turn is responsible for the stem hypertrophy and aerenchyma formation. Adventitious root formation, although a flooding response in some species, is not related to flooding tolerance (12,64).

Another commonly observed response to flooding is stomatal closure. Sometimes it is the earliest observable response (37,64) which is paradoxal because in most species, flooding is usually accompanied by unchanged or increased leaf water potential (12,76). ABA may increase in leaves of flooded plants and is implicated in the closure of the stomata (37,72). In a comparison of *Juglans* species, Shaybany and Martin (72) observed a 6- to 15-fold increase in foliage ABA content within 12-18 h after flooding. This may explain stomatal closure, especially if small, undetectable quantities of ABA are initially released in the vicinity of the guard cells. In this same study, however, ABA concentration declined to control levels within 30 h of flooding in all species except *J. nigra* L. whose leaves desiccated after 24 h. Long term studies have documented stomatal closure for weeks and months, and hence may not involve ABA (37,72).

Bradford (15) suggested an alternative explanation for flooding-induced stomatal closure which involved increased ABA coupled with a simultaneous decrease in cytokinins. Cytokinin production would be affected as rapidly as the observed increases in ACC since both are synthesized in roots. The reduced cytokinin levels in the leaves of flooded *Lycopersicon esculentum* Mill. was found to reduce the regeneration capacity of ribulose bisphosphate, further reducing photosynthesis which already was decreased by stomatal closure (15). Other observations support the role of reduced cytokinin in the flooding response: 1) cytokinin "starvation" leads to chlorosis and senescence of foliage; 2) adventitious root production can be inhibited by high concentrations of cytokinins and gibberellins (64).

2.2. Temperature

2.2.1. *Introduction*. The thermal environment to which trees are exposed varies greatly both diurnally and seasonally. Such fluctuations affect the physiology of the organism, but normally the alterations are reversible, and hence constitute an elastic strain. The most studied of the temperature-related alterations is the development of cold hardiness. This, however, is a special case in which the alterations actually occur in advance of the stress, being triggered by shortening photoperiod and low (above freezing) temperatures. Strictly speaking, cold hardiness is not a physiological response to an adverse temperature stress and hence will not be covered in this chapter.

Trees are exposed to unseasonably high and low temperatures which, depending on duration, severity and stage of tree development, can cause irreversible changes which may be mediated by plant PGRs. As an example, in June 1985 a 2 week period of 35°C and above temperatures stunted newly germinated sand pine (Pinus clausa (Chapm.) Vasey) seedlings

TABLE 1. The proposed role of ethylene in the response of trees to flooding (64).

Physiological Cue	Response
Anaerobiosis in roots	-Activation of ACC synthase in roots
	-Inhibition of ethylene forming enzyme (EFE) in roots leading to accumulation of ACC
	-High ACC content in roots establishes a concentration gradient, initiating loading of ACC into xylem
ACC translocation to shoot	-Movement of ACC to area of higher oxygen at or near water level
	-Rapid conversion of ACC to ethylene by O_2-activated EFE
High ethylene concentration in stem	-Activation of cellulase activity in stem and dissolution of cell walls
	-Cell hypertrophy and aerenchyma formation
	-Adventitious root formation in some species
Increased O_2 in stem and diffusion to roots	-Enhanced O_2 concentration in roots and at least partial amelioration of anaerobiosis

in several pine nurseries located in northern Florida. When normal summer temperatures returned, the shoots never resumed normal growth and they remained stunted for the rest of the growing season. Levitt (44) suggests that most temperature injury, either from high or low temperature, results from direct injury, such as metabolic and membrane disruption. Where PGRs have been observed to change with temperature stress, the significance of those changes is not well understood.

The various organs of a tree have differing sensitivities to temperature stress (43). The roots apparently are more sensitive than stems or foliage, probably because the roots are thermally insulated by soil and do not normally experience the large temperature fluctuations found in the aerial environment. Hence, most information available about temperature stress and PGRs concerns roots.

2.2.2. Low temperature. Low temperatures that are above freezing which cause injury are termed chilling injury (7). The concentration of ABA in root exudates of *Zea mays* varied depending on the temperature to which the root systems were exposed (7). ABA was highest at 13°C and declined with increasing temperature up to 28°C. Root growth was greatest at 23°C and the exudate concentration of cytokinin and gibberellin were highest at 28°C. Membrane damage, as measured by solute leakage from roots, has been associated with plant sensitivity to chilling injury (81). Leakage was prevented if roots were pretreated with ABA, but accelerated if treated with gibberellins (66). In one study, proline concentration increased 20-fold in ABA-treated plants, whereas the control plants only showed a 4-fold increase when chilled (13). Resistance to chilling injury was suggested to be due to the ABA-mediated proline production and the concomitant decrease in osmotic potential (13).

Similar results were reported for Scots pine in a field study comparing a control plot and one in which the soil was artificially cooled (above freezing) over several years (50). Growth was less in the treated plots and was correlated with decreased IAA and GA-like compounds and increased ABA content in the foliage.

2.2.3. High temperature. Like low temperature stress, high temperature stress also leads to changes in PGR levels. Heat shock, two minutes at 47°C, of both *Nicotiana rustica* L. and *Phaseolus vulgaris* L. roots resulted in a halving of the cytokinin concentration and doubling of the ABA concentration (31). Such an extreme stress resulted in a reversal of the trend of decreasing ABA and increasing cytokinin with increasing root temperature noted above by Atkin (7). In cotton (*Gossypium hirsutum* L.) plants grown and presumably acclimated to the high temperature conditions of 42°C, ABA concentration of leaves was slightly less than in control plants grown at 35°C (63). Seedlings of *Taxodium distichum* (L.) Richard and *Nyssa aquatica* L. that had their root systems exposed to 40°C daily for three months exhibited significant reductions in shoot and root biomass, height, and stem diameter. There were no significant changes in photosynthesis, stomatal conductance, transpiration, or root respiration to explain the morphological adaptations (Ken McLeod, unpublished data). Ethylene production by the roots, however, was found to increase slightly in response to the high temperature treatment in *T. distichum* while ethylene production of *N. aquatica* roots was half that of control trees.

2.3. Nutrition

2.3.1. Introduction. Much of the literature dealing with tree nutrition, particularly nutrient deficiencies, has concentrated on description of symptoms and amelioration of the deficiencies with little attention given to the physiology of the tree. In recent years, the effect of deficiencies on carbon partitioning has received increasing attention. Recent reviews of this topic include Ledig (41), Linder and Rook (45) and Cannell (18). It now is commonly accepted that deficiencies will cause a shift in carbon allocation (dry matter production) from shoots to the roots (18,41,45). Research into the questions of if and how PGRs are involved in this stress response is just beginning. The information that is available suggests that PGRs play a key role not only in controlling carbon allocation patterns, but also in other tree responses and in cross-adaptation associated with drought resistance.

2.3.2 Carbon allocation. Ingestad (29) provided an example of the magnitude of change in carbon allocation associated with nutrient stress, particularly nitrogen stress. Birch seedlings grown in low nitrogen partitioned dry matter as follows: leaves, 38%; stem, 7%, and; root, 55%. In contrast, seedlings grown under optimum N conditions partitioned dry matter as follows: leaves, 60%; stem, 20%, and; roots, 20%. Wareing (84) reported a decline in xylem sap cytokinin levels in birch seedlings and sunflower plants when they were transferred to N deficient conditions. Associated with this reduction in cytokinins was the observation of a high degree of apical dominance in the birch seedlings. High N conditions or application of cytokinins to N deficient seedlings results in the loss of apical dominance including the expansion and growth of axillary buds. Dhillon (21) provided collaborative data on cytokinin changes in response to nutrition of sycamore seedlings (*Platanus occidentalis* L.). The highest activity of cytokinin-like material in sap exudate was found in plants growing in the dilution of the Hoagland's solution which produced the maximum dry matter production and relative growth rate. Similar results were obtained when only phosphorus concentrations were varied with the other nutrients remaining at the optimum level. In contrast, a study of P stress in cotton, there was no detectable difference in leaf content of ABA between low and high P treatments (61).

In a series of studies the role of PGRs in controlling carbon partitioning in Scots pine seedlings was investigated. The stimulation of shoot growth in response to increased N was correlated with increases in IAA and free gibberellins, and with decreases in bound gibberellins and ABA-like inhibitors (53). In the roots which exhibited decreased growth in response to the treatment, IAA, cytokinins and ABA-like inhibitors all increased (53). This study was repeated using the Ingestad system (29) for supplying nutrients, and varying light intensities as an additional treatment (51).

The amount of ABA-like inhibitor in needles was found to be lower and in roots to be higher under high N. Light intensity had no effect on the PGR levels. In a third study, potassium nutrition was investigated (52). In the needles, as K increased so did IAA and free gibberellins while ABA and bound gibberellins decreased. With increasing K in the roots, both ABA and free gibberellins increased while bound gibberellins and IAA decreased.

Changes in PGRs in roots and shoots in response to nutrient stress controlling carbon allocation is an attractive hypothesis. The growth promoters (auxins, cytokinins, gibberellins) are increased where carbon is being preferentially allocated while growth inhibitors (ABA) increase where there is a shortage of carbon. These changes in PGRs also influence a tree's response to other stresses.

2.3.3. Stress Interactions. The cross-adaptation of plants appears to hold for the interaction of nutrient stress and water stress. Radin (60) working with nitrogen stress in cotton determined under high water potential the ABA concentration in the leaves to be 8 and 5 ng cm^{-2} for the low and high N treated plants, respectively. This corroborates the data presented in the previous section. The cotton plants were then watered stressed and it was found that the low N plants exhibited ABA-mediated stomatal closure at a leaf water potential of -0.9 MPa whereas the high N plants closed their stomata at -1.7 MPa. Furthermore, for every doubling of the N applied, the water potential threshold for closure decreased by 0.2 MPa. An earlier study established the point of zero turgor in all treatments as -1.5 MPa so the results cannot be explained by increasing osmotic adjustment with increasing N levels (62). Under similar experimental conditions when 10 µM kinetin was applied to low N leaves, stomatal sensitivity to ABA was identical to the high N leaf response (63).

When the effect of P deficiency was investigated in cotton plants subjected to water stress, a similar response was observed (61). Stomatal closure occurred at -1.2 MPa (low P) and -1.6 MPa (high P) with no evidence for a difference in osmotic potentials. At high water potential there was no difference by P treatment in ABA content, but as water stress increased ABA increased more rapidly in the low P plants and led to early stomatal closure. Most of this response could be overcome with exogenous applications of kinetin. Here again evidence suggests that the relative balance between cytokinins and ABA are important in controlling the stress response (stomatal closure) with a modification caused by nutrient stress (30).

2.4. Mechanical

2.4.1. Introduction. Mechanical stresses resulting from wind, rubbing, or flexing can lead to thigmomorphogensis in plants (32). Thigmomorphogenesis is characterized by stunting and stem thickening, and has been associated with reduced photosynthesis and increased respiration (33).

Although not usually considered significant, mechanical stress can severely alter the normal morphological and anatomical development of trees. The extreme manifestation of this stress occurs at treeline where continual high winds cause trees to become procumbant, producing characteristic krummholz. Under less severe conditions, thigmomorphogenesis has been attributed to increasing the wind firmness of stands of trees (77,78).

2.4.2. Thigmomorphogenesis. Early investigations into plant response to mechanical stress found that bending of white pine (*Pinus strobus* L.) branches stimulated ethylene production (42) and that the application of ethephon to branches increased the radial growth of the treated branches (16). The response of bean plants and loblolly pine seedlings to mechanical perturbations was a reduction in stem length and an increase in stem diameter (Table 2). A concomitant increase in ethylene production was observed in both species and the response was found to be mimicked by the application of ethephon (Table 2)(32,33). Thigmomorphogenesis in the two species, however, differed significantly and provided a comparison between a non-woody herbaceous species (bean) and a woody tree species (loblolly pine). As a result of thigmomorphogenesis, the bean plants were found to be more flexible, being able to bend in half without breaking, whereas the pine seedlings became more rigid (33).

Mechanical perturbations, uniformly applied bending, of loblolly pine and frazer fir (*Abies fraseri* (Pursh) Poir.) seedlings caused a peak in ethylene production 18 h after treatment (79). When seedlings of both species were preconditioned to mechanical stress, ethylene production peaked at a lower level after only 8 h in loblolly pine, but

TABLE 2. The role of ethylene in thigmomorphogenesis associated with mechanical stress (32, 33).

Species	Elongation	Stem Thickness	Ethylene Evolution
		Percent of Control	
Bean			
mechanically stressed	86	121	195
control +ethephon	74	158	-
Loblolly Pine			
mechanically stressed	79	109	231
control +ethephon	85	112	-

remained unaltered in fraser fir.

Thigmomorphogenesis has also been demonstrated to induce cross-adaptation to other stresses (33). Bean plants that were mechanically stressed by bending prior to water stress exhibited a 114% increase in dry weight five days after rewatering in comparison to non-stressed plants (33). Likewise, mechanically stressed bean plants showed a 145% increase in height growth in contrast to the controls following a five minute freezing stress at -20°C and four days at 26°C (33). In loblolly pine, seedlings that had been mechanically stressed had 17.5% more live terminals than control seedlings after a freezing stress (33).

Mechanical stress causes the nearly instantaneous increase in callose in the area of the stress (32). Callose, a cell wall polysaccharide, is involved in the activation of the ethylene forming enzyme which catalyzes the conversion of ACC to ethylene and is the controlling factor for the observed ethylene production following mechanical stress (32). Response to mechanical stress is transmitted to adjacent internodes and also across grafts (32). Although the messenger has not been positively identified, both ABA and ACC have been suggested as possible candidates (32).

3. BIOTIC STRESSES

3.1. Disease response and defense

3.1.1. Introduction. Many of the disease symptoms in trees infected with fungi and bacteria are a result of the stress on the tree imposed by the infection (80). The stress symptoms include growth inhibition, epinasty, chlorosis, premature senescence, and leaf abscission (80). Recent studies have implicated PGRs both in the host response in terms of symptom development, and in defense mechanisms of the host. Hoque (27) suggests that PGRs act to control the integration of defense mechanisms in trees and that disease symptoms are related to hormonal imbalances.

3.1.2. Host response. The host tree's response to pathogen infection can be quite rapid and can spread throughout the host. For example, the symptoms listed above are believed to be mediated by ethylene production by the host, pathogen or both (80). Cahill, et al. (17) documented decreases in cytokinins in susceptible *Eucalyptus* species within three days after inoculation of the roots with *Phytophthora*. Both zeatin riboside and isopentenyladenine decreased significantly in the xylem sap prior to the development of any visible symptoms. Such root invading fungi would undoubtedly disrupt normal root physiology, resulting in decreased cytokinin synthesis and translocation. The reduced cytokinin levels may be responsible for the development of the observed symptoms including leaf senescence and abscission.

Changes in cytokinins have also been suggested in the host response to dwarf mistletoe, particularly the formation of witches broom which is a proliferation of lateral shoots in the area of infection (71). Cytokinin assay of shoots of

Arceuthobium vaginatum indicated high quantities of zeatin and zeatin riboside while assays of wood and bark of Pinus ponderosa Laws. infected with the fungi showed the presence of the same cytokinins, but at about one half the concentration found in the shoots (71). An assay of another fungal species, A. occidentalis, which infects ponderosa pine but does not form witches brooms showed low activity of zeatin and high activity in two unidentified cytokinins. With A. vaginatum, the origin of the cytokinin was unknown; it could well have been produced by the fungi and then translocated to the host causing the formation of the witches broom. Regardless of origin, cytokinins are known to overcome apical control of lateral buds and cause the proliferation of lateral shoots (86).

Both cytokinin and IAA levels were found to change in response to a witches broom-forming fungal infection of Paulownia spp (83). Diseased trees had elevated concentrations of both zeatin and isopentenyladenosine while IAA levels were reduced in comparison to healthy trees. In a further attempt to understand the morphogenesis of witches brooms, the proliferation of lateral shoots in vitro of healthy and diseased buds was studied. Lateral shoot development was promoted by high 6 N-benzyl adenine (BA) and low IAA supplied in the culture media to healthy buds. The diseased buds, however, exhibited reduced growth, and lateral shoot production could not be induced even when IAA was increased and BA decreased in the media in an attempt to establish a proper auxin-cytokinin balance for bud growth (83).

3.1.3. Defense mechanisms. Little is known about the role of PGRs in the defense mechanisms of trees. Much of the information available has come from studies with model systems in crop species such as cucumber and bean (27,40). In these systems, ethylene appears to be important in triggering biochemical changes that lead to specific defense responses. Ethylene has been demonstrated to activate phenylalanine ammonia-lyase (PAL), an enzyme of phytoalexin biosynthesis. Phytoalexins are a group of low molecular weight, phenolic compounds that are believed to retard colonization by pathogens (80). ABA is also implicated in the regulation of the biosynthesis of cinnamic acid and phenolic acid, inhibitors which control the level of PAL (27). Ethylene also stimulates other enzyme systems involved in the production of quinones that are toxic to pathogens and the formation of barrier substances such as lignin which wall off the infection point, inhibiting further spread of the pathogen (80).

A paradox, however, exists in this ethylene-mediated defense response. Apparently ethylene of host origin initiates and enhances the mechanisms discussed above, whereas it has been suggested that ethylene of pathogen origin may actually overcome host defenses, acting as a phytotoxin or otherwise modifying tree response (80). In other cases ethylene production increases after visible symptoms have developed, occurring too late to be involved in the initial defense reaction. In these instances it is believed that ethylene production and the associated changes in enzyme activation provide a mechanism to prevent extensive spread of the pathogen from the site of infection (80).

Hoque (27) in a recent review discussed the role of PGRs in the defense mechanism of conifers. In many of the examples cited ethylene played a key role in eliciting defense mechanisms, but he was quick to point out the involvement of the other classes of PGRs. He concluded from a review of the literature that: 1) there was no specificity of stress factors; 2) all classes of PGRs are involved in the stress phenomena; 3) hormonal imbalance is correlated with stress or disease symptoms and; 4) exogenous applications of PGRs can cause similar morphological responses observed under biotic stress.

3.2. Plant-plant communications

3.2.1. Induced resistance. Trees respond to attack by herbivores by increasing the production of phenolic compounds which impart a defense mechanism to the leaves by making them less palatable to the attacking insects (65). Recently, Rhoades (65) found that not only do the attacked trees respond in this manner, but so do surrounding insect-free trees. Changes in leaf quality were induced in red alder (*Alnus rubra* Bong.) and Sitka willow (*Salix sitchensis* Sanson) by insect attacks of tent caterpillars and fall webworms. Attacked trees increased leaf content of proanthocyanidin within 27 days of attack. In Sitka willow both the attacked and unattacked trees showed an altered leaf quality. Under controlled environment, Baldwin and Schultz (9) tore leaves of maple and poplar seedlings in separate studies. Both species responded similarly by increasing total leaf phenolic and tannin content on which the leaves were torn and on undisturbed seedlings growing in the same growth chamber. Control seedlings in a separate growth chamber showed no change in phenolics or tannins.

Although trees under stress emit many volatile compounds, Rhoades (65) identified ethylene as the probable messenger substance that communicates between trees. He cited examples of ethylene-treated bean leaves showing a 50-fold increase in chitinase activity (which is believed to function as an antibiotic) and the ethylene-induced increases of pisatin, isocoumarin and chlorogenic acid in plants. Injured plants possess a systemic alarm signal which may be triggered by ethylene (40). This signal stimulates the accumulation of inhibitors of protease action and does not appear to be host specific (40).

4. SENESCENCE AND ABSCISSION

Stress, both abiotic and biotic, in trees elicits many changes that are primary in nature, that is, the changes in response to the stress enable the trees to better withstand the stress. In addition, there are secondary responses to stress, particularly premature leaf senescence followed by abscission. Both processes have received much attention in the literature (2,35) so they will be only discussed briefly with respect to the role of PGRs in the processes.

The normal ontogeny of trees, especially foliage and fruit, involves tissue senescence. Under stress conditions, these same organs can experience accelerated senescence. In both cases, senescence is characterized by a decline in tissue protein and RNA levels, a loss of chlorophyll, and widespread changes in membrane permeability (8,36). The process of senescence is controlled by a series of changes in the levels of several PGRs (8). One of the first changes is an increased production of ethylene in tissue distal to the area where the abscission zone will develop. IAA levels decrease while ABA levels increase in the abscission zone area, further accelerating tissue senescence by affecting RNA synthesis and stimulating the activity of hydrolytic enzymes (8,35,70). The loss of chlorophyll in leaves marks the latter stages of senescence which is then followed by the process of abscission of the organ.

The role of PGRs in the abscission process is less than clear. Riov et al. (67) suggest that an auxin-ethylene interaction controls the process. Beyer and Morgan (10) postulate that development of an abscission layer involves two steps: 1) a modification of the hormonal balance in the abscission zone with high ethylene levels reducing IAA transport capacity of the petiole and; 2) the direct stimulation of cell wall degrading enzymes by ethylene and the secretion of these enzymes into the cell walls of the abscission zone. An auxin gradient across the abscission zone strongly controls leaf abscission and it is this auxin movement from the leaf to the abscission zone that maintains physiological processes and delays abscission (36).

ABA is also implicated in the abscission process across a wide range of conditions. Addicott (3) contends that ABA controls the abscission zone formation. In this scheme auxin anatagonizes ABA in the process and inhibits abscission. The role of ethylene is diminished, being promotive of abscission but not an absolute requirement for the process to occur (3). The resolution of these conflicting theories awaits additional research.

5. CONCLUSION

Trees throughout their lifetime experience many stresses, sometime singly and sometime in combination. The strain that results from these stresses elicits changes in PGRs, imparting some degree of stress resistance. Looking at the responses to the various stresses, a general pattern becomes evident. In most instances growth promoting PGRs such as auxin and cytokinin decline while growth inhibiting PGRs tend to increase, particularly ABA and ethylene. Little is known about the mode of action of the inhibitors in trees, especially the longer term effects of altered enzyme activity, protein content and RNA turnover. It is also apparent that diverse stresses cause similar plastic strain in trees, lending support to the presence of cross-adaptation in trees.

It should be apparent from this chapter that there are some areas in which research needs to be concentrated including the long-term effects of water stress on tree physiology, the control of carbon partitioning in response to nutrient deficiencies, and the stimulation of biotic stress defense mechanisms.

REFERENCES

1. Abeles FB: Ethylene in Plant Biology. New York: Academic Press, 1973. pp 87-102.
2. Addicott FT: Abscission. Berkeley: University of California Press, 1982.
3. Addicott FT: Abscisis Acid in Abscission. pp. 269-300. In Abscisic Acid, Addicott FT(ed). New York: Praeger Publishers, 1983.
4. Addicott FT and Van Steveninck RFM: Summary: Significance of Abscisic Acid in the Life of Plants. pp 581-586. In Abscisic Acid, Addicott FT(ed). New York: Praeger Publishers, 1983.
5. Amrhein N, Dorzok U, Kionka C, Kondziolka U, Skorupka H, Tophof S: The Biochemistry and Physiology of 1-aminocyclopropane-1-carboxylic Acid Conjugation. pp. 11-20. In Ethylene, Fuchs Y, and Chalutz E(eds). The Hague: Martinus Nijhoff/Dr. W. Junk Publishers, The Netherlands, 1984.
6. Aspinall D: Role of Abscisic Acid and Other Hormones in Adaptation of Plants to Water Stress. pp 155-172. In Adaptation of Plants to Water and High Temperature Stress, Turner NC and Kramer PJ(eds). New York: Wiley, 1980.
7. Atkin RK, Barton GE, Robinson DK: Effect of Root-growing Temperature on Growth Substances in Xylem Exudate of Zea mays. J Exp Bot 24:475-487, 1973.
8. Ayers PG: Water Relations of Diseased Plants. pp. 1-60. In Water Deficits and Plant Growth, Vol V, Kozlowski TT(ed). New York: Academic Press, 1978.
9. Baldwin IT and Schultz JC: Rapid Changes in Tree Leaf Chemistry Induced by Damage: Evidence for Communication between Plants. Sci 221:277-279, 1983.
10. Beyer EM and Morgan PW: Abscission: The Role of Ethylene Modification of Auxin Transport. Plant Physiol 48:208-212, 1971.
11. Blake J and Ferrell WK: The Association Between Soil and Xylem Water Potential, Leaf Resistance, and Abscisic Acid Content in Droughted Seedlings of Douglas-fir (Pseudotsuga menziesii). Physiol Plant 39:106-109, 1977.

12. Blake TJ and Reid DM: Ethylene, Water Relations and Tolerance to Waterlogging of Three Eucalyptus species. Aust J Plant Physiol 8:497-505, 1981.
13. Bornman CH and Jansson E: Nicotiana tabacum callus studies. X. ABA Increases Resistance to Cold Damage. Physiol Plant 48:491-493, 1980.
14. Boussiba S, Rikin A, Richmond AE: The Role of Abscisic Acid in Cross-adaptation of Tobacco Plants. Plant Physiol 56:337-339, 1975.
15. Bradford KJ: Regulation of Shoot Responses to Root Stress by Ethylene, Abscisic Acid, and Cytokinin. pp. 599-608. In Plant Growth Substances 1982, Wareing PF(ed). New York: Academic Press, 1982.
16. Brown KM and Leopold AC: Ethylene and the Regulation of Growth in Pine. Can J For Res 3:143-145, 1973.
17. Cahill D, Grant B, Weste G: How Does Phytophthora cinnamomi Kill a Susceptible Eucalypt? Aust Plant Path 14:59-60, 1985.
18. Cannell MGR: Dry Matter Partitioning in Tree Crops. pp. 160-193. In Attributes of Trees as Crop Plants, Cannell MGR and Jackson JE(eds). Institute of Terrestrial Ecology, 1985.
19. Clemens J and Pearson CJ: The Effect of Waterlogging on the Growth and Ethylene Content of Eucalypus robusta Sm. (Swamp Mahogany). Oecologia 29:249-255, 1977.
20. Davies WJ, Wilson JA, Sharp RE, Osonubi O: Control of Stomatal Behaviour in Water-stressed Plants. pp 163-185. In Stomatal Physiology, Jarvis PG, Mansfield TA(eds). New York: Cambridge University Press, 1981.
21. Dhillon SS: Influence of Mineral Nutrients on Growth and Xylem Sap Cytokinin Level of Sycamore (Platanus occidentalis L.) Seedlings. PhD dissertation, VPI&SU, Dissertation Abstr Intnl B 36:1527B, 1975.
22. Dorffling K, Streich J, Kruse W, Muxfeldt B: Abscisic Acid and After Effect of Water Stress on Stomatal Opening Potential. Z. Pflanzenphysiol 81:43-6, 1977.
23. Glinka Z and Reinhold L: Abscisic Acid Raises the Permeability of Plant Cell Membranes to Water. Plant Physiol 48:103-105, 1971.
24 .Greenwood MS: Reproductive development in loblolly pine. II. The effect of age, gibberellin plus water stress and out-of-phase dormancy on long shoot growth behavior. Amer J Bot 68:1184-1190, 1981.
25 .Greenwood MS and Schmidtling RC: Regulation of catkin production. pp 20-26. In Pollen Management Handbook, Franklin EC(ed). Washinton: USDA For Service Handbook No. 587, 1981.
26. Hale HB: Cross-adaptation. Environ Res 2:423-434, 1969.
27. Hogue E: Biochemical Aspects of Stress Physiology of Plants and Some Considerations of Defense Mechanisms in Conifers. Eur J For Path 12:280-296, 1982.
28. Hoque E, Dathe W, Tesche M, Sembdner G: Abscisic Acid and its β-D-Glucopyranosyl Ester in Saplings of Scots Pine (Pinus sylvestris) in Relation to Water Stress. Biochem Physiol Pflanzen 178:287-295, 1983.

29. Ingestad T and Lund AB: Nitrogen Stress in Birch Seedlings. I. Growth Technique and Growth. Physiol Plant 45:137-148, 1983.
30. Itai C and Benzioni A: Water Stress and Hormonal Response. pp 225-242. In Water and Plant Life, Problems and Modern Approaches, Lange OL, Kappen L, Schulze ED(eds New York: Springer-Verlag, 1976.
31. Itai C, Benzioni A, Ordin L: Correlative Changes in Endogenous Hormone Levels and Shoot Growth Induced by Short Heat Treatments to the Roots. Physiol Plant 29:355-360, 1973.
32. Jaffe MJ: The Involvement of Callose and Elicitors in Ethylene Production Caused by Mechanical Perturbation. pp. 199-215. In Ethylene, Fuchs Y and Chalutz E(eds). The Hague: Martinus Nijhoff/Dr. W. Junk Publishers, 1984.
33. Jaffe MJ and Telewski FW: Thigmomorphogenesis: Callose and Ethylene in the Hardening of Mechanically Stressed Plants. pp. 79-95. In Recent Advances in Phytochemistry, Vol 18, Phytochemical Adaptations to Stress, Timmermann BN, Steelink C, Loewus FA(eds). New York: Plenum Press, 1
34. Johnson JD and Ferrell WK: The Relationship of Abscisic Acid Metabolism to Stomatal Conductance in Douglas-Fir During Water Stress. Physiol Plant 55:431-437, 1982.
35. Kozlowski TT: Tree Growth and Environmental Stresses. Seattle: University of Washington Press, 1979.
36. Kozlowski TT: Shedding of Plant Parts. New York: Academic Press, 1974.
37. Kozlowski TT and Pallardy SG: Effect of Flooding on Water, Carbohydrate, and Mineral Nutrition. pp. 165-194. In Flooding and Plant Growth, Kozlowski TT(ed). Orlando: Academic Press, 1984.
38. Kramer PJ: Plant and Soil Water Relationships: A Modern Synthesis. New York: McGraw-Hill, 1969.
39. Kramer PJ: Water Realtions of Plants. New York: Academic Press, 1983.
40. Kuc' J: Increasing Crop Productivity and Value by Increasing Disease Resistance through Non-genetic Techniques. pp. 147-190. In Forest Potentials, Productivity and Value. Weyerhaeuser Sci Symp, Tacoma, 1985.
41. Ledig FT: The Influence of Genotype and Environment on Dry Matter Distribution in Plants. pp. 427-454. In Plant Research and Agroforestry, Huxley PA(ed). Nairobi International Council fro Research in Agroforestry, 1983.
42. Leopold AC, Brown KM, Emerson FH: Ethylene in the Wood of Stressed Trees. Hortsci 7:175, 1972.
43. Levitt J: Responses of Plants to Environmental Stresses. New York: Academic Press, 1972.
44. Levitt J: Responses of Plants to Environmental Stresses, Vol 1. New York: Academic Press, 1980.
45. Linder S and Rook DA: Effects of Mineral Nutrition on Carbon Exchange and Partitioning in Trees. pp. 211-236. In Nutrition of Forest Trees, Bowen EG and Nambiar KS(eds). New York: Academic Press, 1984.

46. Little CHA: Inhibition of Cambial Activity in Abies balsamea by Internal Water Stress: Role of Abscisic Acid. Can J Bot 53:3041-3050.
47. Loveys BR and Kriedemann PE: Rapid Changes in Abscisic Acid-like Inhibitors Following Alterations in Vine Leaf Water Potential. Physiol Plant 28:476-479, 1973.
48. McKeon TA, Hoffman NE, Yang SF: The Effect of Plant Hormone Pretreatments on Ethylene Production and Synthesis of 1-aminocyclopropane-1-carboxylic Acid in Water Stressed Leaves. Planta 155:437-443, 1982.
49. Markhart AH, Fiscus EL, Naylor AW, Kramer PJ: Effect of Abscisic Acid on Root Hydraulic Conductivity. Plant Physiol 64:611-614, 1979.
50. Menyailo LN, Shul'gina GG, Elagin IN: [Effect of Low Temperature on the Hormone Metabolism of Scots Pine]. Lesovedenie 5:70-74, 1980.
51. Michniewicz M and Stopinska J: The Effect of Nitrogen Nutrition on Growth and on Plant Hormone Content in Scots Pine (Pinus sylvestris L.) Seedlings Grown Under Light of Different Intensity. Acta Soc Bot Pol 49:221-234, 1980.
52. Michniewicz M and Stopinska J: The Effect of Potassium Nutrition on Growth and on Plant Hormone Content in Scots Pine (Pinus sylvestris L.) Seedlings. Ibid 49:235-244, 1980.
53. Michniewicz M, Rozej B, Stopinska J: The Influence of Nitrogen Nutrition on the Dynamics of Growth and Metabolism of Endogenous Growth Regulators in Scotch Pine (Pinus sylvestris L.) Seedlings. Acta Soc Bot Pol 45:495-510, 1976.
54. Milborrow BV: Inhibitors. pp. 76-110. In Advanced Plant Physiology, Wilkins MB(ed). London: Pitman Publishing, 1984.
55. Moore TC: Biochemistry and Physiology of Plant Hormones. New York: Springer-Verlag, 1979.
56. Newville EG and Ferrell WK: Abscisic Acid Levels and Stomatal Behavior During Drought and Recovery in Douglas-fir (Pseudotsuga menziesii). Can J Bot 58:1370-1375, 1980.
57. Pallas JE and Kays SJ: Inhibition of photosynthesis by Ethylene - A Stomatal Effect. Plant Physiol 70:598-601, 1982.
58. Pharis RP and Kuo CG: Physiology of gibberellins in conifers. Can J For Res 7:299, 1977.
59. Pierce M and Raschke K: Correlation Between Loss of Turgor and Accumulation of Abscisic Acid in Detached Leaves. Planta 148:174-182, 1980.
60. Radin JW: Water Relations of Cotton Plants Under Nitrogen Defieciency. III. Stomatal Conductance, Photosynthesis, and Abscisic Acid Accumulation. Plant Physiol 67:115-119, 1981.
61. Radin JW: Stomatal Responses to Water Stress and to Abscisic Acid in Phosphorus-deficient Cotton Plants. Plant Physiol 76:392-394, 1984.

62. Radin JW and Parker LL: Water Relations of Cotton Plants Under Nitrogen Deficiency. II. Environmental Interactions on Stomata. Plant Physiol 64:499-501, 1979.
63. Radin JW, Parker LL, Guinn G: Water Reatlions of Cotton Plants Under Nitrogen Defeiciency. V. Environmental Control of Abscisic Acid Accumulation and Stomatal Sensitivity to Abscisic Acid. Plant Physiol 70:1066-1070, 1982.
64. Reid DM and Bradford KJ: Effects of Flooding on Hormone Relations. pp. 195-221. In Flooding and Plant Growth, Kozlowski TT(ed). Orlando: Academic Press, 1984.
65. Rhoades DF: Pheromonal Communications Between Plants. pp. 195-218. In Recent Advances in Phytochemistry, vol 19, Chemically Mediated Interactions between Plants and Other Organisms, Cooper-Driver GA, Swain T, Conn EE(eds). New York: Plenum Press, 1985.
66. Rikin A, Blumenfeld A, Richmond AE: Chilling Resistance as Affected by Stressing Environments and Abscisic Acid. Bot Gaz 137:307-312, 1976.
67. Riov J, Sagee O, Goren R: Effect of Ethylene on Indole-3-acetic Acid Transport, Metabolism, and Level in Leaf Tissues of Woody Plants During Abscission. pp. 267-276. In Ethylene, Fuchs Y, Chalutz E(eds). The Hague: Martinus Nijhoff/Dr. W. Junk Publishers, 1984.
68. Rose E and Kirkham MB: Genotypic Differences in Ethylene Production by Drought-stressed Wheat Leaves. In Agron Abstr, Ann Meeting , Washington, DC, 1983.
69. Ross SD, Pharis RP, Binder WD: Growth regulators and conifers:Their physiology and potential uses in forestry. pp. 35-78. In Plant Growth Regulating Chemicals, Vol II, Nickell LG(ed). Boca Raton: CRC Press, 1983.
70. Sacher JA: Abscisic Acid in Leaf Senescence. pp. 479-522. In Abscisic Acid, Addicott FT(ed). New York: Praeger Publishers, 1983.
71. Schaffer B, Hawksworth FG, Wullschleger SD, Reid CPP: Cytokinin-like Activity Related to Host Reactions to Dwarf Mistletoes (Arceuthobium spp.). For Sci 29:66-70, 1983.
72. Shaybany B and Martin GC: Abscisic Acid Identification and its Quantitation in Leaves of Juglans seedlings during waterlogging. J Am Soc Hort Sci 102:300-302, 1977.
73. Sivakumaran S, Horgan R, Heald J, Hall MA: Effects of Water Stress on Metabolism of Abscisic Acid in Populus robusta x schied (P. 'robusta') and Euphorbia lathyrus L.. Plant, Cell and Enivron 3:163-173, 1980.
74. Stumpff NJ: Ethylene Production by Loblolly Pine Seedlings During Cold Storage and Water Stress. MS Thesis, VPI&SU, Blacksburg, Virginia, 1984.
75. Stumpff NJ and Johnson JD: Ethylene Production by Loblolly Pine Seedlings Associated With Water Stress. Physiol Plant (in Press), 1896.
76. Tang ZC and Kozlowski TT: Some Physiological and Morphological Responses of Quercus macrocarpa seedlings to flooding. Can J For Res 12:196-202, 1982.

77. Telewski FW and Jaffe MJ: Thigmomorphogenesis: Field and Laboratory Studies of Abies fraseri in Response to Wind or Mechanical Perturbation. Physiol Plant 66:211-218, 1986.
78. Telewski FW and Jaffe MJ: Thigmomorphogenesis: Anatomical, Morphological and mechanical Analysis of Genetically Different Sibs of Pinus taeda in Response to Mechanical Perturbation. Ibid 66:219-226, 1986.
79. Telewski FW and Jaffe MJ: Thigmomorphogenesis: The Role of Ethylene in the Response of Pinus taeda and Abies fraseri to Mechanical Perturbation. Ibid 66:227-233, 1986.
80. Van Loon LC: Regulation of Pathogens and Symptom Expression in Diseased Plants by Ethylene. pp. 171-180. In Ethylene, Fuchs y and Chalutz E(eds). The Hague: Martinus Nijhoff/Dr. W. Junk Publishers, 1984.
81. Van Steveninck RFM and Van Steveninck ME: Abscisic Acid and Membrane Transport. pp. 171-236. In Abscisic Acid, Addicott FT(ed). New York: Praeger Publishers, 1983.
82. Walton DC: Biochemistry and Physiology of Abscisic Acid. Ann Rev Plant Physiol 31:453-489, 1980.
83. Wang R, Wang S, Sun X:[Effects of Hormones on the Morphogenesis of Witches Broom of Paulownia spp.]. Scientia Silvae Sinicae 17:281-286, 1981.
84. Wareing PF: Growth Substances and Integration in the Whole Plant. Symp Soc Exp Biol 31:337-365, 1977.
85. Wright STC: The Relationship Between Leaf Water Potential and the Levels of Abscisic Acid and Ethylene in Excised Wheat Leaves. Planta 134:183-189, 1979.
86. Zimmerman MH and Brown CL: Trees: Structure and Function. New York: Springer-Verlag, 1971.

Plant Growth Regulation 6: 217–236

10. THE ROLE OF PLANT GROWTH REGULATORS IN THE DEVELOPMENT AND GERMINATION OF CONIFER POLLEN

J.E. WEBBER
RESEARCH SCIENTIST, BRITISH COLUMBIA MINISTRY OF FORESTS AND LANDS, RESEARCH BRANCH, 1320 GLYN ROAD, VICTORIA, B.C., V8Z 3A6, CANADA

ABSTRACT

All classes of PGRs have been identified in pollen and at least implicated in its growth and development. When added exogenously, their effects on in vitro pollen germination and tube growth can be either neutral, promotive or inhibitory. It is generally agreed that PGRs play an important role in regulating germination and tube growth but endogenous levels appear to be sufficient and significant responses to in vitro applications are seldom observed. Specific roles of PGRs in conifer pollen have not been elucitated but certain speculations are offered. In pines, auxins derived from pollen may be important in the continued development of the ovule, since unpollinated ovules abort. However, in several other genera (Pseudotsuga for example) pollen is not required to continue ovule development. Gibberellins (GAs) may have a role in the early stages of germination and tube growth. Both polar and non-polar GAs have been isolated from conifer pollen but in germinating pollen, it is the polar GAs which rise and the non-polar GAs that decline during the time of maximum tube growth. Abscisic acid (ABA) occurs in the pollen, styles and ovaries of many angiosperm species and some conifers but it does not have any apparent role in germination or tube growth. It may, however, play an important role in regulating incompatibility reactions. Brassinolides have not been studied in any detail but their occurrence in conifer pollen has been confirmed. Cyclic-adenosine monophosphate (cAMP) may not be considered a PGR but it has been identified in conifer pollen and it does mimic the effects of other PGRs in pollen germination studies. In vitro PGR effects are discussed in relationship to media effects which can have a dramatic influence on pollen germination and tube growth and may explain the lack of a PGR response in certain cases.

1. INTRODUCTION

Phytoactive substances isolated from pollen and other floral tissues and their effects on the reproductive processes in plants represent some of the earliest work in plant growth regulator (PGR) research. Compared to the multicellular sporophyte, pollen is a relatively simple system and conifer pollen in particular, can be handled, stored and cultured in vitro with moderate ease. All classes of PGRs have been either identified or implicated in the development, germination and tube growth of conifer pollen. However, their specific function is largely unknown.

Access to abundant supplies of conifer pollen with a genetic heritage is becoming more available as tree improvement programs are established. Furthermore, physiological and environmental restraints affecting flowering and the production of orchard seed have been identified (49,87,90) and in certain cases largely overcome (93). Interest in conifer pollen and the dynamic role it plays in the transfer of the male gamete in forest tree seed orchards is now receiving the attention of forest research and management programs.

Most seed orchards rely on pollen to be freely available and to mate equally with all parent trees. Unrestricted random mating between all orchard parent trees forms the basis of potential genetic gain calculations for progeny. The degree to which improved orchard seed approaches this theoretical gain calculation has been termed genetic efficiency (1,35). Realizing full genetic efficiency from orchard seed is predicated on several assumptions, most of which are seldom attained. For example, deviation from random mating caused by asynchronous receptivity periods and unequal fecundity between orchard parent trees, inbreeding, and pollen contamination occur in various degrees within most orchards (23,36,38,121).

Various operational techniques are being employed to improve genetic efficiency (34). One technique, supplemental mass pollination (SMP), has received considerable attention for improving both genetic efficiency and seed yields (13,36,119).To be effective, however, SMP must be accompanied with a good pollen management program (117) based on a thorough knowledge of the species pollination mechanism (87). Even when good pollen handling technique is followed, there is no guarantee that SMP will improve efficiences or yields (88,118).

Use of SMP to manipulate the parental base of an orchard requires pollen of the highest possible quality with a vigor at least equal to naturally shed pollen. There are stages within the pollination and fertilization process where male/male competition and male/female interactions can occur which ma have special implications for seed production. For example, there are some studies with angiosperm pollen which indicate that gametophytic selection and/or competition may affect sporophytic traits (69,85,94,124). Within conifers there is an indication that pollen competition based on viability can occur (118) but its effect on progeny performance is not known. The possiblity of affecting gamete traits or at least imparting a competitive advantage to applied pollen over undesirable pollen could be an important seed orchard management tool. In this respect, the role played by growth regulators in affecting development, germination and fertilization of conifer pollen is of interest.

2. POLLEN DEVELOPMENT

The response of pollen germination and tube growth to PGRs is most often studied _in vitro_. The sensitivity of these _in vitro_ assays to PGRs can be expected to vary depending on the maturation state and vigor of the pollen. It is important then, to understand the various patterns of pollen-cone development and recognize the sensitive stages for each species.

Development of pollen-cone buds (microsporangiate strobilii) is similar for most north temperate conifers although considerable variation exists between genera regarding the stage of development reached when winter dormancy occurs (86). All species initiate pollen sacs (microsporangia) on leaf-like structures (microsporophylls) before winter dormancy. For most species, dormancy is followed by a period of rapid development including meiotic divisions of pollen mother cells to form microspores, cell division and maturation of the pollen grain. In _Chamaecyparis_ and _Juniperus_ species, anatomical maturation of pollen grains is complete before winter dormancy.

The meiotic process is a very sensitive developmental stage for all pollen including conifers. Extreme temperatures (4,37,100) and water stress (17) during meiosis have been associated with increased meiotic irregularities resulting in poor pollen viability or vigor.

Several examples of abnormal pollen development in angiosperms arising from adverse environmental conditions during meiosis have been reported. In wheat, male sterility was induced by both high temperatures (97) and water stress (95) applied during meiosis of the pollen mother cells. Female sterility was also induced but only by high temperatures.

Since abscisic acid (ABA) accumulates in various plant organs under water stress (116), its possible role in inducing male sterility was tested. Sani and Aspinall (96) found that ABA (3×10^{-5} M) applied just before meiosis of the pollen mother cells mimiced the effect of water deficit resulting in male sterility. There was no effect on female fertility, however, and a post-meiotic application of ABA had no effect on either male or female fertility. Sani et al. (98) investigated the anatomical events leading to pollen abortion in response to heat, water deficit and ABA application and concluded that ABA did not play a primary role in water-stress induced male sterility in wheat. An association between ABA content and fertility in the wheat spike was demonstrated by Zeng and King (125). Male-sterile spikes from a chromosomal mutant had an ABA content that was significantly higher and a rate of ABA degradation that was almost one-half that of a normal cultivar.

Pollen sterility in wheat has also been induced by ethrel application (3000 ppm) without affecting female fertility (31). As for ABA, the maximum gametocidal response of ethrel occurred when plants were treated immediately prior to meiosis of the pollen mother cells (31 and references therein). Emasculation and pollen sterility have also been observed with the application of ethrel to Indian mustard (5) and morphactin to Lagenaria siceraria (105).

3. POLLINATION

Although there is no reported role of PGRs in conifer pollination, there are several prezygotic stages where pollen can interact with the female tissue and PGR effects are possible. Conifer pollen relies on the random process of wind pollination and success depends on several factors not least of which is the availability of a large number of pollen grains (120).

3.1 Pollination Mechanisms

There are four basic mechanisms by which the functional ovuliferous scales capture pollen (86). The most familiar pollination mechanism is exemplified by Pinus and Picea species where a pollination drop is exuded through the micropylar opening between the integument arms. Adhesion of pollen landing on the edge of the micropyle or on the integument arms is aided by the secretion of minute droplets which form before the pollination drop appears. As the pollination drop recedes, pollen is taken into the micropyle.

Tsuga and Abies species are examples of two other mechanisms where the integument tip is a simple funnel-shaped opening and no pollination drop occurs. Originally, it was thought that only pollen landing either in the funnel or on the edge of the funnel rim could be taken into the micropylar canal when the integument tip collapsed. This apppears to be true for Abies. However, Colangeli and Owens (20) have recently shown that Tsuga heterophylla pollen actually germinates outside the micropyle forming a very long pollen tube which by some unknown process, is directed into the micropyle and to the nucellus. Only a few other conifers display a pollination mechanism similar to this (see 87).

Pseudotsuga and Larix represent the fourth type of pollination mechanism. In these genera, the ovule tips develop into two unequal lobes covered by unicellular stigmatic hairs leaving a narrow slit-like micropyle. Pollen grains landing on these hairs become entrapped and after several days beyond seed-cone bud burst these hairs begin to collapse and the cells around the micropyle elongate engulfing any trapped pollen within the micropyle.

3.2 Pollen Rehydration

Pollen is normally shed in a desiccated state (between 15-30% moisture content) and depending on the species and the extent of desiccation, viability can be adversely affected. Angiosperm pollen is more sensitive to desiccation, and a reduction of moisture content to 20% or less is associated with a drastic reduction in viability, especially vigor (7,42). Conifers, however, are far less sensitive to desiccation and can be maintained in the dehydrated state for several years (117).

Pollen germination is usually preceded by rehydration which in most cases is a prerequisite step (42). Shivanna and Heslop-Harrison (103) report that structural changes occur to the pollen membrane during dehydration and that viability of the pollen is determined by the stability of the membrane in this altered state. If membrane reorganization occurring during dehydration is reversible, then pollen grains, under suitable conditions, can recover and produce normal pollen tubes.

Seed orchards often require stored pollen for future use and a reduction of moisture content to less than 10% is usually required for successful storage (117). However, Charpentier and Bonnet-Masimbert (18) have shown that reducing pollen moisture content to below 8% causes a significant reduction of in vitro germination in Douglas-fir pollen. A rehydration treatment at 100% RH for 16 h, however, improved germination percentage.

Dehydration of Douglas-fir pollen does not appear to affect in vivo fertilization, however. Mellerowicz and Bonnet-Masimbert (73) were not able to show any significant difference between dry or rehydrated Douglas-fir pollen in cone development or seed yields. Apparently, the rehydration period in this conifer is less urgent, probably because in vivo germination does not begin until several days after pollination (see 87).

Important steps which may be affected by plant growth regulators are pollen germination and tube growth, pollen/nucellus interaction, pollen tube/megagametophyte interaction and embryo development. The remainder this review will consider the effects of growth regulators on pollen germination and tube growth and briefly discuss the pollen/nucellus interaction.

4. OCCURRENCE OF PGRS IN CONIFER POLLEN

Table 1 lists the studies on conifer pollen which have either indicated growth regulator activity or attempted to identify specific compounds. It is worth noting that nearly all studies reviewed have dealt with only one genus, Pinus.

Auxins, gibberellins (GAs), ABA and inhibitors have been identified in pine pollen and the presence of cytokinins suggested. Although ethylene probably occurs, it has not, to my knowledge, been identified in conifer pollen. Also, brassinolides, which occur in the pollen of several angiosperms (9) have also been identified in P. densiflora pollen (123).

TABLE 1: IDENTIFIED PLANT GROWTH REGULATORS IN THE POLLEN OF PINUS

PGR Class by Species	Activity	Assay	(REF)
Auxins			
P. sylvestris	IAA, IAN		(74)
P. radiata	IAA	Avena coleoptile	(108, 109)
P. sibirica	Indolic		(62)
Cytokinins			
P. radiata	CK-like	carrot tissue culture	(109)
Inhibitors			
P. radiata	3 inhibitors	Avena coleoptile	(109)
P. sylvestris			(74)
P. sibirica	phenolic		(62)
P. denisflora			(112, 113)
" "			(55)
" "	ABA		(102)
Gibberellins			
P. radiata	4 GA-like	radish cotyledon barley endosperm rumex leaf disc	(108, 109)
P. sylvestris	GA-like		(74)
P. attenuata	GA_3, GA_4, GA_9	GC-MS	(50)
P. attenuata[1]	GA-like	dwarf rice	(53)
P. coulteri	- polar	cv. Tan-genbou	
P. ponderosa	- less polar		
P. attenuata	GA_1, GA_{34}	GC-tritiated assay	(52)
P. sylvestris	GA-like		(48)
	GA-like		(57)
Pseudotsuga menziesii	GA_3	gas chromatography	(21)
	4 GA-like	chromatography	(21)
Brassinosteriods			
P. thubergii	2-deoxycastasterone castasterone		(123)
Cylic-3',5' Phosphoric Acids			
P. densiflora	cAMP cGMP		(111)

1 Assays of ungerminated and germinating pollen

Inhibitors, auxin- and GA-like activity were detected in ungerminated *P. sylvestris* pollen (74). Using paper chromatography and the *Avena* coleoptile bioassay, Michalski (74) observed a promoter (possibly indole acetonitrile) and inhibitors in the neutral-ether fraction; two promoters (one of which co-chromatographed with IAA) and two inhibitors in the acid-ether fraction; and three zones of GA-like activity in the acid-ether fraction.

Tanaka (112) separated the acidic-ether extract of both germinated and ungerminated pollen of *P. densiflora* by paper chromatography. Several compounds were detected which inhibited the growth of pollen tubes in *Tradescantia*, *Impatiens* and *Lilium* as well as pine. Tanaka (113) also showed that as germination of the pine pollen progressed, the activity of the inhibitor extract decreased. The inhibitor fraction was not identified but Tanaka (113) showed evidence that it behaved like a phenolic compound.

In germinated and ungerminated *P. radiata* pollen, Sweet and Lewis (108,109) detected three inhibitors, five auxin-like, and four GA-like compounds and one fraction with both GA-like and cytokinin-like activity. One promoter, showing auxin activity, diffused out of the germinating pollen into an agar substrate. Sweet and Lewis (108) suggested that this auxin diffuses into the nucellus where it affected continued ovule development (in pine, unpollinated ovules abort). As *in vitro* pollen germination progressed, Sweet and Lewis (109) bioassayed fractions from either paper or thin-layer chromatograms using six systems. The quantities of some substances increased, others decreased and still others remained constant. One promoter which behaved like an auxin increased in activity as pollen germination proceeded over time. A second promoter from the aqueous extract, believed to be a gibberellin, decreased to no detectable activity after 72 h suggesting to these authors that it had a possible role in either germination or early tube growth of *P. radiata* pollen.

In *Pseudotsuga menziesii*, Crozier *et al.* (21) identified GA_3 by column and gas chromatography as a major GA component of pollen in addition to four other GA-like fractions. Endogenous GAs have been identified by bioassay in *P. attenuata*, *P. coultrei* and *P. ponderosa* pollen (53). Bioassay analysis of gibberellin fractions obtained from column chromatography during a 72 h germination period showed a wide range of GA activity ranging from non-polar to highly polar. Maximum GA activity, corresponding to the elution zone of GA_3 was observed after 15 h of germination. Thereafter, all GA activity declined to no detectable amounts at 72 h. Kamienska and Pharis (53) suggested that GAs are utilized in pollen tube growth since the maximum activity (mostly in the polar region) at 15 h corresponded to the period of most rapid pollen tube growth.

In a more detailed analysis of *P. attenuata* pollen by gas chromatography-mass spectrometry, Kamienska *et al.* (50) identified GA_3, GA_4, and GA_7 and characterized a fourth compound similar to GA_9. The concentrations of GA_9-like, GA_4 and GA_7 all decreased during germination, whereas the peak GA activity in the more polar region increased. The most predominant polar peak after 15 h germination was GA_3.

It appears that GA biosynthesis in higher plants proceeds from the less polar to the more oxygenated, less saturated and hence more polar GAs (32). In germinating *P. attenuata* pollen (52), GA_1 and GA_{34} (polar GAs) were found as products of metabolism of 1,2-[^{3}H]GA_4 (less polar). Both polar and non-polar GAs are present in ungerminated pollen but upon germination, the concentration of polar GAs increases, at the apparent expense of less polar GAs. Pinus species differ from most other conifers in requiring an extra year to complete fertilization. Futhermore, in vitro pollen tube growth in pines typically slows down after 72 h. What then, in the in vivo reproductive system of pine is made available to continue pollen tube growth, arrest it for a year and then stimulate its growth again to enable penetration of the nucellus and fertilization to occur?

As discussed earlier, Tanaka (112) demonstrated the presence of inhibitors in P. densiflora pollen and their inhibitory effect on pollen germination and tube growth. Kobayashi et al. (55) identified and characterized benzoic acid, p-hydroxybenzoic acid and dehydrochorismic acid from P. densiflora pollen and related benzoic acid structural types to inhibitory activity. However, relatively large concentrations of these compounds (500 mg L^{-1}) were required before significant reduction in either germination or tube growth was observed. The inhibitory activity of these acids all increased as the pH was lowered from 7 to 4.

ABA has been isolated from P. densiflora pollen (102) but it was not related to inhibition of pollen germination or tube growth. ABA has been identified in the styles, ovaries, anthers and pollen of higher plants and its possible role in regulating incompatibility reactions will be discussed later.

5. PGR EFFECTS ON POLLEN GERMINATION AND TUBE GROWTH

Some of the earliest observations in PGR studies were that both pollen and floral tissues contained active substances capable of affecting pollen germination and tube growth, fertilization, and ovule development (see 107). Malik and Bhandal (64) have reviewed the effects of PGRs on pollen germination and tube growth applicable to all species including some conifers. This section will be concerned with PGR effects on conifer pollen only.

Kuhlwein (61) found extracts of stigmas and pollen in Cypripedium species improved germination of P. montana pollen and tube growth in Abies species. He further suggested that the pollination drop in Pinus and Picea species contained growth hormones capable of enhancing pollen germination. McWilliam (71) argued that the stimulation of pollen germination and tube growth in pine was not due to the presence of growth regulators in the pollination drop but rather to the amino acid and sugar (sucrose, glucose, and fructose) constituents. McWilliam (72) also found that P. resinosa pollen would not germinate in its own pollination drop and concluded that this was not the germination medium for pine.

5.1 Media Effects

Media effects have by far the most important influence on pollen germination and tube growth. Most in vitro germination assays assume that the number of pollen grains producing tubes in a given time (usually less than 48 h) is an index of viability. However, pollen germinated in vitro seldom attains the size or rate of tube growth observed in vivo.

Factors affecting in vitro pollen germination include pH, temperature, oxygen uptake, medium constituents (inorganics and carbohydrates), and osmotic pressure (80,81,107). The two most consistently important medium constituents for most angiosperm pollen and many conifers are boron (50,115) and calcium (11).

For conifers, medium requirements vary for species and successful germination conditions range from plain water to complex aqueous or agar solutions containing various salts and sugar components. When an attempt is made to optimize medium constituents, increases in germination, tube growth and metabolic activity are normally observed (41,80,81,82,84). The presence of a sugar for short term germination periods is often not required. However, its presence for longer germination periods (>48 h) is required as source of energy and carbon for continued biosynthesis and growth, and perhaps as important, as an osmoticum to stabilize the medium and to prevent pollen grains from bursting (84).

For P. roxburghii pollen, Dhawan and Malik (25) demonstrated the essential requirement of sucrose in the media before germination could occur. Addition of boric acid to the sucrose-medium further enhanced germination and tube growth, although boric acid by itself had no effect. Nygaard (81) found that P. mugo pollen could germinate in distilled water alone but tube growth was improved by the addition of sugars and salts including calcium and boric acid. Calcium, magnesium, boric acid and nitrate all were essential and the stimulation due to calcium required either magnesium or boric acid. Apparently, calcium is essential to maintain the structural integrity of the membrane (81,28) whereas the role of borate in pollen germination and tube growth is not known. Possibly borate facilitates the uptake of sugars from the germination media (115) or it may affect growth-related processes within the cell or transport processes at the cell surface (30).

5.2 PGR Effects on Conifer Pollen Germination

The few studies which have assessed the PGRs effects on in vitro germination of conifer pollen are shown in Table 2.

TABLE 2. EFFECTS OF EXOGENOUS PLANT GROWTH REGULATORS ON IN VITRO GERMINATION IN SOME CONIFERS.

Species (Ref)	PGR	Effect[1]	
		G	T
Pinus mugo (80)	IAA		ne[2]
	GA_3		ne
	Kinetin		ne
	2,4-D		-
	ABA		ne
P. roxburghii (56)	IAA	-	
	Kinetin	-	+
P. roxburghii (25)	IAA	+	+
	GA_3	+	+
	Ethylene	+	+
	ABA	+	+
	cAMP	+	+
Pinus ellioti (33)	IAA	-	
	2,4-D	-	
P. radiata (109)	IAA	-	
	2,4-D	-	
P. densiflora (111)	cAMP		-
	cGMP		+
Pseudotsuga menziesii (19)	GA_1	+	
(76)	IAA		
	GA_1		-
	$GA_{4/7}$		-
	Kinetin		-

[1] G = % germination
T = Tube growth

[2] ne : no effect
- : inhibiting effect
+ : enhanced effect

Konar (56) observed a stimulation of P. roxburghii pollen tube growth with IAA (2.5 mg L^{-1}) and kinetin (100 mg L^{-1}). Dhawan and Malik (25) found similar results for P. roxburghii but also observed an inhibiting effect of IAA at higher concentrations (50 mg L^{-1}). Furthermore, Dhawan and Malik (25) demonstrated a promotive effect on germination and tube growth by GA_3, ethylene, ABA and cyclic AMP at concentrations between 1 and 10 mg L^{-1}. Sweet and Lewis (109) found that kinetin (10 mg L^{-1}) enhanced tube growth in P. radiata pollen whereas GA_3 had no effect at concentrations below 0.01 mg L^{-1}. At concentrations above 0.01 mg L^{-1}, GA_3 was significantly inhibitory. In P. mugo, IAA, GA_3, kinetin, 2,4-D and ABA (all at concentrations between 0.1 and 10 mg L^{-1}) had no effect on percent germination and tube growth (80). In P. elliotii, however, Echols and Mergen (33) found that IAA at concentrations of 1 to 10 mg L^{-1} completely inhibited pollen germination. In Pseudotsuga menziesii, GA_3 at concentrations between 10 and 1000 mg L^{-1} significantly enhanced tube growth and advanced cytological development (19). However, Muren et al. (76) found that GAs (GA_1 and $GA_{4/7}$), IAA, and kinetin at similar concentrations added either singly or in combination to the germination media of P. menziesii did not stimulate pollen elongation or gamete development.

These studies show that reported effects of PGRs on in vitro germination and tube growth are inconsistent. Much of this variation can possibly be related to differences in media and germination conditions, in the age of pollen (i.e., fresh or stored), pollen storage conditions and pollen genotypes. Presumably there are interactions between different media constituents such that growth regulator effects may vary depending on which component is limiting.

5.3 PGR Effects on Angiosperm Pollen Germination

Effects of PGRs on in vitro germination and tube growth in angiosperms have been studied in more detail. Most authors generally agree that PGRs play an important role in mediating germination and tube growth (15,25,65,67,70,89, 92). However, specific roles of individual regulators have not been elucidated.

When added exogenously, IAA has both inhibited (2) and slightly promoted (10,22,91) germination and tube growth. In Zea mays, Dhingra and Varghese (26) found IAA did not affect pollen germination in plants grown under normal culture. However, in plants raised under stress conditions induced by various levels of salinity, IAA at lower concentrations (<10 mg L^{-1}) significantly enhanced tube growth (26). In a number of other plants, Loo and Hwang (63) reported various degrees of enhanced pollen germination and tube growth from exogenous IAA. Using the data reported by Loo and Hwang (63), Nissen (78) found, in most cases, the response of IAA was subsensitive. This classication of IAA response was based on whether the change from 10 to 90% of maximal response required less than, as much as, or more than, an 81-fold increase in external concentration. The response was then classified as ultrasensitive, hyperbolic or subsensitive, respectively (78).

Exogenous GAs affect germination and tube growth as well. As for IAA, in vitro response of GAs vary considerably with species (16), by genotype (89) and as a function of their concentration (15,89).

6. ROLE OF PGRS IN POLLEN GERMINATION

6.1 Auxins

The presence of auxins in pollen (see 6), their production in the style and ovary associated with pollen tube growth and fertilization, and the resultant stimulation of ovary growth following fertilization are well established (79). In addition to auxins, gibberellins and cytokinins are also associated with fruit set. Nagar and Raja Rao (77) detected low quantities of auxins, gibberellins and cytokinins in unpollinated ovaries of guava fruits (Psidium guajava L.) at anthesis but in pollinated ovaries the activity of these PGRs increased while in unpollinated ovaries their activity decreased.

6.2 Gibberellins

As previously mentioned, GAs occur in both ungerminated and germinated pine pollen (see Table 1) and as germination progresses, their level varies both qualitatively and quantitatively (6,53,107). Chandler (16) considered GAs to be important in the initial stages of germination. Based on a sharp drop in GA content during the first hour of germination in Lilium henryi pollen, Barendse et al. (6) also concluded that GAs have an important role in the initial stages of germination. It appears that most pollen contains adequate levels of GAs (and auxins) which may explain why in vitro application of exogenous PGRs have had little or no effect on pollen germination and tube growth (30,80).

6.3 Abscisic Acid

Several studies (62,109,112) report the presence of inhibitors in pollen, some of which have been identified as substituted benzoic acid derivatives (55). In P. densiflora pollen, the activity of these soluble inhibitors decline upon germination suggesting a possible role in the initiation of germination (113).

Shibuya et al. (102) have isolated ABA from P. densiflora pollen, but Sondheimer and Linskens (106) did not consider ABA (nor IAA, zeatin, or GA_3) to play a major role in regulating pollen germination or tube growth (in Petunia hybrida at least). When pollen was derived from corn plants grown under salt stress, exogenous ABA caused a reduction in germination response (26). However, for those pollen grains which did germinate, ABA promoted tube growth (26).

In Amaryllis pollen, Sidhu et al. (104) observed a promotive affect of ABA in the concentration range of 1 to 10 mg L^{-1} on pollen tube elongation. ABA also enhanced the non-photosynthetic fixation of $^{14}CO_2$, and the activity of phosphoenolpyruvate carboxylase during in vitro germination. At a concentration of 1 mg L^{-1}, IAA, GA_3, N6-benzyladenine and ethephon had no effect on pollen tube elongation but at higher concentrations (10 mg L^{-1}), their effect was inhibitory.

6.4 Cyclic-3',5' Phosphoric Acid

Whether cyclic-adenosine monphosphate (cAMP) has a regulatory role in plants is debatable (3). Both cAMP and cyclic-guanosine monophosphate (cGMP) have been isolated from P. densiflora pollen and their levels increase as germination progresses (111). Applied exogenously, cAMP slightly inhibited and cGMP slightly enhanced in vitro pollen tube elongation in P. densiflora pollen. In Tradescantia paludosa (66) and P. roxburghii (25) pollen, exogenous cAMP enhanced tube elongation. Addition of cAMP (1 or 10 mg L^{-1}) to IAA, GA, ethylene, or ABA had an additive effect on promoting germination and tube growth in P. roxburghii pollen which suggested to Dhawan and Malik (25) that cAMP mimics the effect of other PGRs.

Katsumata et al. (54) proposed that cAMP has a role in the initiation of pollen germination in P. densiflora pollen. It was observed that levels of cAMP and adenlate cyclase activity increased during pollen germination and continued to increase even when germination was inhibited with cycloheximide. According to Malik et al. (66), cAMP increases the imbibition of water, activates protein synthesis and hastens the metabolic processes leading to tube formation, including biogenesis of wall polysaccharides and increased respiration. Dhawan and Malik (24) observed enhanced activity of peroxidases, malate dehydrogenase and succinic dehydrogenase due to cAMP, suggesting tube growth enhancement is caused by activating the respiratory cycle.

7. BIOCHEMICAL FACTORS ASSOCIATED WITH IN VITRO GERMINATION

The inconsistency of exogenous growth regulator response on in vitro pollen germination and tube growth may be the result of an insensitive system (114). As previously mentioned, membrane structure can alter dramatically with pollen dehydration, and restoring viability is dependent on the pollen's ability to reorganize its membranes (103). Futhermore, pollen viability can be expected to vary with genotype, handling technique and cultural manipulation. Their is some information at the biochemcial level which may explain variation in viability and perhaps explain some of tne observed inconsistencies in growth regulator response. There is, however, little information on conifer pollen and much of our working knowledge comes from angiosperm systems.

Brewbaker and Majumder (12) have compared pollen of different angiosperm families and classified them as either binucleate (generative cell undivided at anthesis) or trinucleate (pollen contains two sperm nuclei at anthesis). Compared to trinucleate pollen, binucleate pollen is much easier to store and readily germinates in culture. At maturity, the metabolic activity of trinucleate pollen (and some binucleate species as well) occur at a rapidly respiring state. Less developed binucleate pollen, however, are characterized by slower respiration rates, prolonged lag periods prior to germination, and a complete inhibition of protein synthesis and tube emergence with cycloheximide (43,44). In many regards, conifer pollen can be considered as similar to this latter group.

Both germination and early tube elongation in binucleate pollen are independent of RNA synthesis (68), and this appears to be true for pine pollen as well (83). In apple pollen, early tube elongation was independent of RNA synthesis (110), but after a certain period (about 3 h) tube growth could be retarded with actinomycin D, indicating a requirement for new RNA synthesis. Calzoni and Speranza (14) also found that germination and early tube growth in apple pollen were not affected by either actinomycin D or cycloheximide. The later stages of tube growth, however, required new RNA and protein synthesis.

Tagliasacchi et al. (110) described these two distinct phases of pollen germination and tube growth (which occur both in vitro and in vivo) as autotrophic and heterotrophic. In the first, autotrophic phase pollen germination and early tube growth is derived from stored material. In vitro, this phase can last from just a few hours (3 h for apple) up to 12 h for pine (80). The second, heterotrophic phase, requires substances from the growth medium or female tissue to continue and, in general tube growth accelerates.

During a 4-day germination period of Pseudotsuga menziesii pollen, respiration rates remained constant during the first 48 h in culture, then increased to four times the original value (76). Starch content decreased fom an original value of 30% to about 10% and the respiratory quotient remained relatively constant reflecting carbohydrate utilization. It is possible then that the autotrophic phase in P. menziesii pollen can be extended because of the high initial starch content. During this same period, soluble protein, insoluble protein, and RNA content decreased by about 50%, whereas DNA quantity doubled (76). The adenine nucleotide content remained constant during the germination period but Muren et al. (76) noted a rapid increase in ATP content during the first 8 h which remained high for the remainder of the germination period.

In mature apple pollen, starch content is low but during in vitro germination starch begins to accummulate just before tube emergence and is hydrolyzed when tube growth ends (8). The period of rapid starch synthesis occurring after the onset of germination in lily pollen also coincides with a period of initially high respiration (29). This is followed by a short period of low respiration when tube emergence occurs, and a third phase (heterotrophic?) of rapid tube growth when respiration rates are highest (27,29).

There are a few studies which have looked at plant growth regulators in relationship to the time course of pollen germination. Within hours of initiating germination, the pollen of Petunia and Lilium exhibited a sharp drop in GA activity (6). A similar decline in GAs occured for three species of Pinus (53). Maximum GA activity occurred after 15 h of incubation, which corresponded with the onset of rapid pollen tube growth. During this period, the less polar GAs declined, presumably being converted to more polar GAs (50). Metabolic turnover of GAs was rapid and [^{3}H]-GA_4 (a less polar GA) was converted to the more polar GA_1, GA_2 and the biologically inactive GA_{34} (52). This initial turnover of less polar GAs to the more active, polar GAs coincided with the early stages of tube emergence.

8. COMPATIBLITY

In many angiosperm species, a self-incompatibility mechanism causes the rejection of pollen from the same or closely related plants. This pollen-stigma and/or pollen-style interaction is determined by one or more self-incompatible (S) loci. An incompatibility reaction occurs when the S allele products in the pistil and pollen have the same specificity, but other mechanisms have been proposed as well (see 47).

Although there are no comparable pollen-pistil interactions in gymnosperms, incompatibility systems do exist (39). Apparently the maternal sporophyte of conifers lacks the ability to recognize pollen (58,99). However, interspecific incompatibility in the form of inhibited tube growth through the nucellus tissue has been observed in Picea (59,75) and Pinus (40,60).

Plant growth regulators have been implicated in incompatibility reactions in angiosperms. Pollen germination inhibitors have been detected in Brassica stigmas pollinated with incompatible self-pollen but have not been found in stigmas pollinated with compatible outcross pollen (45,46). The nature of these inhibitors has not been determined, but Sedgley (101) detected flavanoids within the pollen and stigmas of B. oleracea. Hodgkin and Lyon (45) have suggested that some of these compounds may play a role in preventing germination and tube growth in incompatible pollinations.

In Petunia hybrida, ABA content of styles increased rapidly during floral development but remained relatively constant in the ovaries and anthers (122). After compatible cross-pollination, the ABA content of the styles in particular, but also of the ovaries, was significantly reduced compared with incompatible self-pollination. ABA content in the pollen of Lilium longiflorum decreased upon germination, while at the same time a large increase in ABA content of the germination medium was observed (122). The authors were not prepared to speculate on whether the build up of ABA in pollen and its subsequent release upon germination played a role in regulating growth of the pollen tube.

9. CONCLUSIONS

Results from studies on both angiosperm and conifer pollen suggests that PGRs are likely involved in all reproductive steps from initiation to fertilization. There are no definitive studies which demonstrate a specific role of PGRs in conifer reproduction. However, their presence in pollen has been confirmed and their activity at certain stages has at least implicated PGRs in germination and tube growth. It is also likely that PGRs play an important role in the continued development of the ovule.

Auxins are known for their role in stimulating ovary development in angiosperms and it is possible they play a similar role in conifers. Certainly without pollen, pine strobili will abort but in other conifers (i.e. Pseudotsuga and Tsuga species), normal cone and ovule development can occur in the absence of pollen. Whether the nature of the stimuli arising from pine pollen is an auxin or some other factor remains to be elucidated. Auxins have also been implicated in stimulating pollen germination but in general in vitro applications of IAA have shown either a weak stimulatory or neutral effect. If IAA is involved in the germination process, in vivo levels must be sufficient to meet this requirement.

Gibberellins appear to play an important role in the early stages of pine pollen germination. Metabolic activity favors the production of more polar GAs which at their peak period of activity, correspond to the maximum rate of pollen tube extension. However, all GA activity declines to undectable levels after about 72 h. Since germination occurs over a considerably longer period in vivo, it is likely the female tissue supplies other sustaining factors. What interaction between the male and female gametophyte exists is not known but it appears that the initial requirements for pollen germination are at least carried by the pollen. This is perhaps one reason why GAs, like auxins, show an inconsistent response to in vitro application.

Several inhibitors including ABA have been isolated from conifer pollen but there is no known function for any of them in the germination process. A role of ABA in incompatiblity reactions in angiosperms seems possible but there is no evidence a similar role occurs in conifers. If ABA were to affect pollen germination in vivo, it would most likely be involved in the pollen/nucellus interaction. As for auxins and GAs, ABA shows no consistent effect to in vitro germination and tube growth.

Either PGR levels in pollen do not limit in vitro germination or our ability to observe their response has not been optimized. When an attempt is made to maximize pollen germination and tube growth in culture, the effects of media constituents become very important. In fact, the prerequisite of hyration prior to germination and the requirement of boron, calcium and a sugar source in the medium have a far more dramatic effect on pollen germination and tube growth than do PGRs. As we improve our ability to culture pollen in vitro, the analysis of PGRs may become more meaningful. Furthermore, as more genotypes become available from tree improvement programs, it is possible that pollen types with limiting levels of PGRs can be identified and used for furthering our understanding of their role in the reproductive process.

REFERENCES

1. Adams WT (1983) Application of isozymes in tree breeding. In:Tanksley SD and Orton TJ, eds. Isozymes in Plant Genetics and Breeding, Part A, pp 381-400. Amsterdam: Elsevier Sci Pub
2. Addicott FT (1943) Pollen germination and pollen tube growth as influenced by pure plant growth substances. Plant Physiol 18:270-279
3. Armhein N (1974) Evidence against the occurrence of cAMP in higher plants. Planta 118:240-258
4. Andersson E (1980) Temperature conditioned irregularities in pollen mother cells of Picea abies (L.) Karst. Hereditas 92:27-35
5. Banga SS and Labana KS (1984) Ethrel induced male sterility in Indian mustard (Brassica juncea (L) Coss). Z Pflanzenzuchtg 92:229-233
6. Barendse GWM, Pereira ASR, Berkers PA, Driessen FM, van Eyden-Emons and Linskens HF (1970) Growth hormones in pollen, styles and ovaries of Petunia hybrida and of Lilium species. Acta Bot Neerl 19:175-186
7. Barnabas B (1985) Effect of water loss on germination ability of maize (Zea mays L.) pollen. Ann Bot 55:201-204
8. Bellani LM, Pacini E and Franchi GG (1985) In vitro pollen germination and starch content in species with different reproductive cycle. II. Malus domestica Borkh. cultivars starkcrimson and golden delicious. Acta Bot Neerl 34:65-71
9. Bonetti A, Cerana R, Lado P, Marre E, Marre MT and Romani G (1983) Mechanism of action of the pollen hormone brassinolide. In: Mulcahy DL and Ottaviano E, eds. Pollen: Biology and Implications for Plant breeding, pp 9-14. New York: Elsevier Sci Pub
10. Bose N (1959) Effect of gibberellic acid on the growth of pollen tubes. Nature (Lond) 184:1577
11. Brewbaker JL and Kwack BH (1963) The essential role of calcium ion in pollen germination and pollen tube growth. Amer J Bot 50:859-865
12. Brewbaker JL and Majumder SK (1961) Cultural studies of the pollen population effect and the self-incompatibility inhibition. Amer J Bot 48:457-464
13. Bridgwater FE and Trew IF (1981) Supplemental mass pollination. In: Franklin EC, ed. Pollen Management Handbook, pp 52-57. USDA For Ser Agric Hndbk No 587
14. Calzoni GL and Speranza A (1986) Developmental stage of mature apple pollen: study of RNA and protein synthesis during in vitro germination. Physiol Veg 24:53-62
15. Carmichael JW (1970) The effect of gibberellic acid on in vitro pollen germination in Digitaria pentzii Stent. Proc Soil Crop Sci Soc Florida 30:255-258
16. Chandler C (1957) The effect of gibberellic acid on germination and pollen tube growth. Contrib Boyce Thompson Inst 19:215-244
17. Chandler C and Mavrodineanu S (1965) Meiosis of Larix laricina Koch. Contrib Boyce Thompson Inst 23:67-75
18. Charpentier JP and Bonnet-Masimbert M (1983) Effect of prior rehydration on the germination in vitro of Douglas-fir (Pseudotsuga menziesii) pollen after storage. Ann Sci For 40:309-317
19. Ching KK and Ching TM (1959) Extracting Douglas-fir pollen and effects of gibberellic acid on its germination. For Sci 5:74-80
20. Colangeli A and Owens JN (1984) Pollination mechanism in western hemlock. In: Proc West For Gen Assoc Ann Meet. Genetic gain: Incorporation and Implication for Forestry, p 20. Victoria: Univ of Victoria

21. Crozier A, Kuo CC, Durely RC and Pharis RP (1970) The biological activties of 26 gibberellins in nine plant bioassays. Can J Bot 48:867-877
22. de Bruyn JA (1966) The in vitro germination of pollen of Setaria sphacelata. I. Effects of carbohydrates, hormones, vitamins and micronutrients. Physiol Plant 19:365-376
23. Denison NP and Franklin EC (1975) Pollen management. In: Faulkner R, ed. Seed Orchards, pp 92-100. London: For Comm Bull No 54
24. Dhawan AK and Malik CP (1979) Cyclic-AMP control of some oxido-reductases during pine pollen germination and tube growth. Phytochem 18:2015-2017
25. Dhawan AK and Malik CP (1981) Effect of growth regulators and light on pollen germination and pollen tube growth in Pinus roxburghii Sarg. Ann Bot 47:239-248
26. Dhingra HR and Varghese TM (1985) Effect of growth regulators on the in vitro germination and tube growth of maize (Zea mays L.) pollen from plants raised under sodium chloride salinity. New Phytol 100:563-569
27. Dickinson DB (1965) Germination of lily pollen: respiration and tube growth. Science 150:1818-1819
28. Dickinson DB (1967) Permeability and respiratory properties of germinating pollen. Physiol Plant 20:118-127
29. Dickinson DB (1968) Rapid starch synthesis associated with increasing respiration in germinating lily pollen. Plant Physiol 43:1-8
30. Dickinson DB (1978) Influence of borate and pentaerythitol concentrations on germination and tube growth of Lilium longiflorum pollen. J Amer Soc Hort Sci 103:413-416
31. Dotlacil L and Apltauerova M (1978) Pollen sterility induced by ethrel and its utilization in hybridization of wheat. Euphytica 27:353-360
32. Durley RC and Pharis RP (1973) Interconversion of gibberellin A_4 to gibberellin A_1 and A_{34} by dwarf rice cultivar Tan-ginbozu. Planta b:357-361
33. Echols RM and Mergen F (1956) Germination of slash pine pollen in vitro. For Sci 2:321-327
34. El-Kassaby YA (1986) CIP Forest Products' tree improvement program and forest genetics activities 1983-1985. In: Yeatman CW and Boyle TJB, eds. Proc 20th Can Tree Impr Assoc Meet, Part I, pp 231-236. Ottawa: Can For Ser
35. El-Kassaby YA, Fashler AMK and Sziklai O (1984) Reproductive phenology and its impact on genetically improved seed production in a Douglas-fir seed orchard. Silvae Genet 33:120-125
36. El-Kassaby YA and Ritland K (1986) The relation of outcrossing and contamination to reproductive phenology and supplemental pollination in a Douglas-fir seed orchard. Silvae Genet 35:240-244
37. Eriksson G (1968) Temperature response of pollen mother cells in Larix and its importance for pollen formation. Stud For Succ No 63:1-131
38. Eriksson G, Jonsson A and Lindgren D (1973) Flowering in a clone trial of Picea abies Karst. Stud For Succ No 110:1-45
39. Hagman M (1975) Incompatibility in forest trees. Proc R Soc Lond B 188:313-326
40. Hagman M and Mikkola L (1963) Observations on cross-, self- and interspecific pollinations in Pinus peuce Grisbe. Silvae Genet 12:72-79
41. Hellmers H and Machlis L (1956) Exogenous substrate utilization and fermentation by the pollen of Pinus ponderosa. Plant Physiol 31:284-289
43. Heslop-Harrison J (1979) An interpretation of the hydrodynamics of pollen. Amer J Bot 66:737-743

43. Hoesktra FA (1983) Physiological evolution in angiosperm pollen: possible role of vigour. In: Mulcahy DL and Ottaviano E, eds. Pollen: Biology and Implications for Plant Breeding, pp 35-41. New York: Elsevier Sci Pub
44. Hoekstra FA and Bruinsma J (1979) Protein synthesis of binucleate and trinucleate pollen and its relationship to tube emergence growth. Planta 146:559-566
45. Hodgkin T and Lyon GD (1984) Pollen germination inhibitors in extracts of *Brassica oleracea* L. stigmas. New Phytol 96:293-298
46. Hodgkin T and Lyon GD (1986) The effect of *Brassica oleracea* stigma extracts on the germination of *B. oleracea* pollen in a thin layer chromatographic bioassay. J Exp Bot 37:406-411
47. Hogenboom NG (1975) Incompatibility and incongruity: two different mechanisms for the non-functioning of intimate partner relationships. Proc R Soc Lond B 188:361-375
48. Ivonis I Yu (1969) Gibberellin-like substances in young cones and pollen of Scotch pine trees. Fiziol Rast 16:937-939
49. Jackson DI and Sweet GB (1972) Flower initiation in temperate woody plants. Horticult Abstr 42:9-24
50. Johri BM and Vasil IK (1961) Physiology of pollen. Bot Rev 27:325-381
51. Kamienska A, Durley RC and Pharis RP (1976a) Isolation of gibberellins A_3, A_4 and A_7 from *Pinus attenuata* pollen. Phytochem 15:421-424
52. Kamieńska A, Durley RC and Pharis RP (1976b) Endogenous gibberellins of pine pollen. III. Conversions of 1,2-[^{3}H]GA_4 to gibberellins A_1 and A_{34} in germinating pollen of *Pinus attenuata* Lemm. Plant Physiol 58:68-70
53. Kamienska A and Pharis RP (1975) Endogenous gibberellins of pine pollen. II. Changes during germination of *Pinus attenuata*, *P. coulteri* and *P. ponderosa* pollen. Plant Physiol 56:655-659
54. Katsumata T, Takahashi N and Ejiri SI (1978) Biochemical studies on pollen. XXI. Changes of cyclic AMP level and adenylate cyclase activity during germination of pine pollen. Agric Biol Chem 42:2161-2162
55. Kobayashi S, Ariga M, Ozawa T and H Imagawa (1984) Pollen growth inhibitors in *Pinus densiflora* pollen: isolation, biological activities, and structure-activity relationship. Agric Biol Chem 48:389-395
56. Konar RN (1958) Effect of IAA and kinetin on the pollen tube growth of *Pinus roxburghii* Sar. Cur Sci 27:216-217
57. Kopcewicz J (1969) The dynamics of growth substances during the development of the pine inflorescences. I. Male inflorescences. Roczn Nauk Roln 95(Ser A):105-112
58. Koski V (1973) On self-pollination, genetic load and subsequent inbreeding in some conifers. Comm Inst For Fenn 78:1-42
59. Kossuth SV and Fechner GH (1973) Incompatibility between *Picea pungens* Engelm. and *Picea engelmannii* Parry. For Sci 19:50-60
60. Kriebel HB (1972) Embryo development and hybridity barriers in white pines (section Strobus). Silvae Genet 21:39-44
61. Kuhlwein H (1948) Uber keimungsfordernde substanzen in pollen und narben. Planta 35:528-535
62. Larionova NA, Minina EG, Kir'yanova IA, Tolkachev ON and Mitrofanova TK (1977) Auxins, growth inhibitors, and lipids of the pollen of siberian pine. Fiziol Rast 24:175-179
63. Loo TL and Hwang TC (1944) Growth stimulation by manganese sulphate, indole-3-acetic acid, and colchicine in pollen germination and pollen tube growth. Amer J Bot 31:356-367

64. Malik CP and Bhandal IS (1983) Hormonal regulation of pollen germination and pollen tube growth. In: Purohit SS, ed. Aspects of Physiology and Biochemistry of Plant Hormones, pp 217-253. New Dehli, Kalyani
65. Malik CP and Chhabra N (1976) Hormonal regulation of pollen germination and tube elongation in Arachis hypogaea. Proc Ind Acad Sci 84B:101-b
66. Malik CP, Chhabra N and Vermani S (1976) Cyclic AMP-induced elongation of pollen tubes in Tradescantia paludosa. Biochem Physiol Pflanzen 169S:311-315
67. Malik CP and Mehan M (1975) Correlative effects of auxins, gibberellic acid, kinetin on the elongation of pollen tubes in Calotropis procera. Biochem Physiol Pflanzen 167:295-301
68. Mascarenhas JP and Bell E (1969) Protein synthesis during germination of pollen. Studies on polyribosome formation. Biochem Biophys Acta 179:199-203
69. McKenna MA and Mulcahy DL (1983) Ecological aspects of gametophytic competition in Dianthus chinensis. In: Mulcahy DL and Ottaviano E eds. Pollen: Biology and Implications for Plant Breeding, pp 419-424. New York: Elsevier Sci Pub
70. McLeod KA (1975) The control of growth of tomato pollen. Ann Bot 39:591-596
71. McWilliam JR (1958) The role of the micropyle in the pollination of Pinus. Bot Gaz 120:109-117
72. McWilliam JR (1960) Pollen germination of Pinus as affected by the environment. For Sci 6:27-39
73. Mellerowicz E and Bonnet-Masimbert M (1986) The importance of moisture content of pollen used in controlled crosses for Douglas-fir. Ann Sci For 43:179-188
74. Michalski L (1967) Growth regulators in the pollen of pine (Pinus sylvestris L.). Acta Soc Bot Pol 36:475-481
75. Mikkola L (1969) Observations on interspecific sterility in Picea. Ann Bot Fenn 6:285-339
76. Muren RC, Ching TM and Ching KK (1979) Metabolic study of Douglas-fir pollen germination in vitro. Physiol Plant 46:287-292
77. Nagar PK and Raja Rao T (1986) Early changes in growth regulator content of pollinated guava fruits. Scientia Hortic 29:139-146
78. Nissen P (1985) Dose responses of auxins. Physiol Plant 65:357-374
79. Nitsch JP (1952) Plant hormones in the development of fruits. Quart Rev Biol 27:33-57
80. Nygaard P (1969) Studies on the germination of pine pollen (Pinus mugo) in vitro. I. Growth conditions and effects of pH and temperature on germination, tube growth and respiration. Physiol Plant 22:338-346
81. Nygaard P (1970) Studies on the germination of pine pollen (Pinus mugo) in vitro. II. Effects of different ions. Physiol Plant 23:372-384
82. Nygaard P (1971) Studies on the germination of pine pollen (Pinus mugo) in vitro. III. Inhibition by d-mannose and deoxyhexoses. Physiol Plant 24:130-135
83. Nygaard P (1973) Nucleotide metabolism during pine pollen germination. Physiol Plant 28:361-371
84. Nygaard P (1977) Utilization of exogenous carbohydrates for tube growth and starch synthesis in pine pollen suspension cultures. Physiol Plant 39:206-210
85. Ottaviano E, Sair-Gorla M and Mulcahy DL (1980) Pollen tube growth rate in Zea mays: implications for genetic improvement of crops. Science 210:437-438

86. Owens JN (1980) Reproductive biology of conifers. In: Dancik BP and Higginbotham KO eds. Proc 6th North Amer For Biol Workshop, pp 77-92. Edmonton: Univ of Alberta
87. Owens JN and Blake MD (1985) Forest tree seed production. Can For Ser Info Rep PI-X-53:1-161
88. Owens JN and Simpson SJ (1982) Further observations on the pollination mechanism and seed production of Douglas-fir. Can J For Res 12:431-434
89. Pfahler PL, Wilcox M, Mulcahy DL and Knauft DA (1982) In vitro germination and pollen tube growth of maize (Zea mays L.) pollen. X. Pollen source genotype and gibberellin A_3 interactions. Acta Bot Neerl 31:105-111
90. Puritch GS (1972) Cone production in conifers. Can For Ser Info Rep BC-X-65:1-39
91. Raghavan V and Baruah HK (1959) Effect of time factor on the stimulation of pollen germination and pollen tube growth by certain auxins, vitamins, and trace elements. Physiol Plant 12:441-451
92. Rosen WG (1968) Ultrastructure and physiology of pollen. Ann Rev Plant Physiol 19:435-462
93. Ross SD and Pharis RP (1986) Status of flowering in conifers: a constraint to tree improvement? In: Caron F, Corriveau AG and Boyle TJB, eds. Proc 20th Can Tree Impr Assoc Meet, Part 2, pp 11-28. Ottawa: Can For Ser
94. Sacher RF, Mulcahy DL and Staples RC (1983) Developmental selection during self pollination of Lycopersicon x solanum F1 for salt tolerance of F2. In: Mulcahy DL and Ottaviano E, eds. Pollen: Biology and Implications for Plant Breeding, pp 329-334. New York: Elsevier Sci Pub
95. Sani HS and Aspinall D (1981) Effect of water deficit on sporogenesis in wheat (Triticum aestivum L.). Ann Bot (Lond) 48:623-633
96. Sani HS and Aspinall D (1982) Sterility in wheat (Triticum aestivum L.) induced by water deficit or high temperature: possible mediation by abscisic acid. Aust J Plant Physiol 9:529-537
97. Sani HS, Sedgley M and Aspinall D (1983) Effect of heat stress during floral development on pollen tube growth and ovary anatomy in wheat (Triticum aestivum L.). Aust J Plant Physiol 10:137-144
98. Sani HS, Sedgley M and Aspinall D (1984) Developmental anatomy in wheat of male sterility induced by heat stress, water deficit or abscisic acid. Aust J Plant Physiol 11:243-253
99. Sarvas R (1968) Investigation on the flowering and seed crop of Pinus sylvestris. Comm Inst For Fenn 53:1-198
100. Sarvas R (1972) Investigations on the annual cycle of development of forest trees: active period. Comm Inst For Fenn 76:1-110
101. Sedgley M (1975) Flavanoids in pollen and stigma of Brassica oleracea and their effects on pollen germination in vitro. Ann Bot 39:1091-1095
102. Shibuya T, Funamizu M and Kitahara Y (1978) Abscisic acid from Pinus densiflora pollen. Phytochem 17:322-323
103. Shivanna KR and Heslop-Harrison J (1981) Membrane state and pollen viability. Ann Bot 47:759-770
104. Sidhu RK, Basra AS and Malik CP (1986) Hormonal effects on tube elongation, $^{14}CO_2$ fixation and phosphoenolpyruvate carboxylase activity in Amaryllis pollen: promotion by abscisic acid. Plant Growth Reg 4:293-298
105. Singh KP, Singh I and Kaushik MP (1983) Growth, flowering, sex behaviour, pollen sterility and fruit set in Lagenaria siceraria Standle as as affected by morphactin. Ind J Plant Physiol 36:15-20

106. Sondheimer E and Linskens HF (1974) Control of in vitro germination and tube extension of Petunia hybrida pollen. In: Proc Kon Nederl Akad Wet Ser C, 77:116-124
107. Stanley RG and Linskens HF (1974) Pollen: Biology Biochemistry and Management. Springer-Verlag New York. 307p
108. Sweet GB and Lewis PN (1969) A diffusible auxin from Pinus radiata pollen and its possible role in stimulting ovule development. Planta (Berl) 89:380-384
109. Sweet GB and Lewis PN (1971) Plant growth substances in pollen of Pinus radiata at different levels of germination. N Z J Bot 9:146-156
110. Tagliasacchi AM, Forino IMC, Bellani LB and Avanzi S (1985) Dynamics of pollen grain germination in two cultivars of Malus domestica: the influence of actinomycin D on fresh and stored pollen grains. Ann Bot 56:29-33
111. Takahashi N, Ejiri S and Katsumata T (1978) Changes of cyclic AMP and cyclic GMP levels during germination of pine pollen. Agric Biol Chem 42:1605-1606
112. Tanaka A (1958) The pollen germination and pollen tube development in Pinus densiflora Sieb. et Zucc. III. The growth inhibiting substances in the ether extract from Pinus pollen grains. Sci Rep Tohoku Univ Ser 4 (Biol) 24:45-54
113. Tanaka A (1964) The pollen germination and pollen tube development in Pinus densiflora Sieb. et Zucc. VII. The changes is the amount of acidic growth inhibitors during pollen germination. Sci Rep Tohoku Univ Ser 4 (Biol) 30:211-217
114. Trewavas AJ (1982) Growth substance sensitivity: the limiting factor in plant development. Physiol Plant 55:60-72
115. Vasil IK (1964) Effect of boron on pollen germination and pollen tube growth. In: Linskens HF ed. Pollen Physiology and Fertilization, pp 107-119. Amsterdam: North-Holland Pub Co
116. Walton DC (1980) Biochemistry and physiology of abscisic acid. Ann Rev Plant Physiol 31:453-489
117. Webber JE (1987) Increasing seed yield and genetic efficiency in Douglas-fir seed orchards through pollen management. For Ecol Man (in press)
118. Webber JE and Yeh FCH (1987) Test of the first-on first-in pollination hypothesis in coastal Douglas-fir. Can J For Res 17:63-68
119. Wheeler NC and Jech KS (1986) Estimating supplemental mass pollination (SMP) success electophoretically. In: Caron F, Corriveau AG and Boyle TJB, eds. Proc. 20th Can Tree Imp Assoc Meet, Part 2, pp 111-120. Ottawa: Can For Ser
120. Whitehead DR (1983) Wind pollination: some ecological and evolutionary perspectives. In: Real L, ed. Pollination Biology, pp 97-108. New York: Academic Press Inc
121. Woessner RA and Franklin EC (1973) Continued reliance on wind-pollinated southern pine seed orchards - is it reasonable? In: Proc 12th South For Tree Imp Con, pp 64-73. Baton Rouge: Louisiana State Univ
122. Yang ZH, Barendse GWM and Linskens HF (1985) Abscisic acid in the reproductive organs of Petunia hybrida and Lilium longiflorum. Acta Bot Neerl 34:73-82
123. Yokota T, Arima M, Takahashi N, Takatsuto S, Ikekawa N and Takematsu T (1983) 2-Deoxycastasterone, a new brassinolide-related bioactive steroid from Pinus pollen. Agric Biol Chem 47:2419-2420

124. Zamir D, Tanksley SD and Jones RA (1981) Low temperature effect on selective fertilization by pollen mixtures of wild and cultivated tomato species. Theor Appl Genet 59: 235-238
125. Zeng ZR and King RW (1986) Regulation of grain number in wheat: changes in endogenous levels of abscisic acid. Aust J Plant Physiol 13:347-352

Index

Hormonal Control of Tree Growth

Chapter 1

Chapter 2

Chapter 3

Chapter 4

Chapter 5

Chapter 6

Chapter 7

Chapter 8

Chapter 9

Chapter 10

www.ingramcontent.com/pod-product-compliance
Ingram Content Group UK Ltd.
Pitfield, Milton Keynes, MK11 3LW, UK
UKHW020806290926
15470UKWH00003B/988
9789024736218